Electricity and Magnetism

by Elisha Gray

CONTENTS.

CHAPTER

INTRODUCTION.

For the benefit of the readers of Vol. III, who have not read the general Introduction found in Vol. I, a word as to the scope and object of this volume will not be amiss.

It will be plain to any one on seeing the size of the little book that it cannot be an exhaustive treatise on a subject so large as that of Electricity. This volume, like the others, is intended for popular reading, and technical terms are avoided as far as possible, or when used clearly explained. The subject is treated historically, theoretically, and practically.

As the author has lived through the period during which the science of Electricity has had most of its growth, he naturally and necessarily deals somewhat in reminiscence. All he hopes to do is to plant a few seed-thoughts in the minds of his readers that will awaken an interest in the study of natural science; and especially in its most fascinating branch--Electricity.

If Vol. I is at hand, please read the Introduction. It will bring you into closer sympathy with the author and his mode of treatment.

Again, if the reader is especially interested in the theory of Electricity it will help him very much if he first reads Vols. I and II, as a preparation for a better understanding of Vol. III. All the natural sciences are so closely related that it is difficult to get a clear insight into any one of them without at least a general idea of all the others.

NATURE'S MIRACLES.

ELECTRICITY AND MAGNETISM.

CHAPTER I.

THE AUTHOR'S DESIGN.

The writer has spent much of his time for thirty-five years in the study of electricity and in inventing appliances for purposes of transmitting intelligence electrically between distant points, and is perhaps more familiar with the phenomena of electricity than with those of any other branch of physics; yet he finds it still the most difficult of all the natural sciences to explain. To give any satisfactory theory as to its place with and relation to other forms of energy is a perplexing problem.

It is said that Lord Kelvin lately made the statement that no advance had been made in explaining the real nature of electricity for fifty years. While this statement--if he really made it--is rather broad, it must be acknowledged that all the theories so far advanced are little better than guesses. But there is value in guessing, for one man's guess may lead to another that is better, and, as it is rarely the case that each one does not give us a little different view of the matter, it may be that out of the multiplicity of guesses there may some time be a suggestion given to some investigator that will solve the problem, or at least carry the theme farther back and establish its true relationship to the other forms of energy. I cannot but think that there is yet a simple statement to be made of Energy in its relation to Matter that will establish a closer relationship between the different branches of physical science. And this, most likely, will be brought about by a better understanding of the nature of the interstellar substance called Ether, and its relation to all forms and conditions of sensible matter and energy.

In the talks that will follow it will be the endeavor of the writer to give such a simple and popular exposition of the phenomena and applications of electricity, in a general way only, that the popular reader may get, at least, an elementary understanding of the subject so far as it is known. As we have said, the descriptions will have to be elementary, for nothing else can be done without such elaborate technical drawings and specifications as would be impossible in our limited space, and would not be clear to the ordinary reader who knows nothing of the science.

Thousands who are employed in various ways with enterprises, the foundations of which are electrical, know nothing of electricity as a science. A friend of mine, who is a professor of physics in one of our colleges, was traveling a few years ago, and in his wanderings he came across some sort of a factory where an electric motor was employed. Being on the alert for information, he stepped in and introduced himself to the engineer, and began asking him questions about the electric motor of which he had charge. The professor could talk ohm, amperes, and volts smoothly, and he "fired" some of these electrotechnical names at the engineer. The engineer looked at him blankly and said: "You can't prove it by me. I don't know what you're talking about. All I know is to turn on the juice and let her buzz." How much "juice" is wasted in this cut-and-dry world of ours and how much could be saved if only all were even fairly intelligent regarding the laws of nature! A great deal of the business of this world is run on the "let her buzz" theory, and the public pays for the waste. It will continue to be so until a higher order of intelligence is more generally diffused among the people. A fountain can rise no higher than its source. A business will never exceed the intelligence that is put into it, nor will a government ever be greater than its people.

Let us begin the subject of electricity by going somewhat into its past history. It is always well to know the history of any subject we are studying, for we often profit as much by the mistakes of others as by their successes. I shall also give the theories advanced by different investigators, and if I should have any thoughts of my own on the subject I shall be free to give them, for I have just as good a right to make a guess as any one. It must be confessed, however, that the older I grow the less I feel that I know about the subject of electricity, or anything else, in comparison with what I see there is yet to be known. I once met a young man who had just graduated from college, and in his conversation he stated that he had taken a course in electricity. I asked him how long he had studied the subject. He said "three months." I asked him if he understood it--and he said that he did. I told him that he was the man that the world was looking for; that I had studied it for thirty years and did not understand it yet.

"A little learning is a dangerous thing"--for it puffs us up, and we feel that we know it all and have the world in our grasp; but after we have tried our "little learning" on the world for a while and have received the many hard knocks that are sure to come, we are sooner or later brought up in front of the mirror of experience, and we "see ourselves as others see us," and are not satisfied with the view.

Whatever the theories may be regarding electricity, and however unsatisfactory they may be, there are certain well-defined facts and phenomena that are of the greatest importance to the world. These we may understand: and to this end let us especially direct our efforts.

CHAPTER II.

HISTORY OF ELECTRICAL SCIENCE.

Electricity as a well-developed science is not old. Those of us who have lived fifty years have seen nearly all its development so far as it has been applied to useful purposes, and those who have lived over twenty-five years have seen the major portion of its development.

Thales of Miletus, 600 B.C., discovered, or at least described, the properties of amber when rubbed, showing that it had the power to attract and repel light substances, such as straws, dry leaves, etc. And from the Greek word for amber--elektron--came the name electricity, denoting this peculiar property. Theophrastus and Pliny made the same observations; the former about 321 B.C., and the latter about 70 A.D. It is also said that the ancients had observed the effects of animal electricity, such as that of the fish called the torpedo. Pliny and Aristotle both speak of its power to paralyze the feet of men and animals, and to first benumb the fish which it then preyed upon. It is also recorded that a freed-man of Tiberius was cured of the gout by the shocks of the torpedo. It is further recorded that Wolimer, the King of the Goths, was able to emit sparks from his body.

Coming down to more modern times--A.D. 1600--we find Dr. Gilbert, an Englishman, taking up the investigation of the electrical properties of various substances when submitted to friction, and formulating them in the order of their importance. In these experiments we have the beginnings of what has since developed into a great science. He made the discovery that when the air was dry he could soon electrify the substances rubbed, but when it was damp it took much longer and sometimes he failed altogether. In 1705

Francis Hawksbee, an experimental philosopher, discovered that mercury could be rendered luminous by agitating it in an exhausted receiver. (It is a question whether this phenomenon should not be classed with that of phosphorescence rather than electricity.) The number of investigators was so great that all of them cannot be mentioned. It often happens that those who do really most for a science are never known to fame. A number of people will make small contributions till the structure has by degrees assumed large proportions, when finally some one comes along and puts a gilded dome on it and the whole structure takes his name. This is eminently true of many of the more important developments in the science and applications of electricity during the last twenty-five or thirty years.

Following Hawksbee may be mentioned Stephen Gray, Sir Isaac Newton, Dr. Wall, M. Dupay and others. Dupay discovered the two conditions of electrical excitation known now as positive and negative conditions. In 1745 the Leyden jar was invented. It takes its name from the city of Leyden, where its use was first discovered. It is a glass jar, coated inside and out with tin-foil. The inside coating is connected with a brass knob at the top, through which it can be charged with electricity. The inner and outer coatings must not be continuous but insulated from each other. The author's name is not known, but it is said that three different persons invented it independently, to wit, a monk by the name of Kleist, a man by the name of Cuneus, and Professor Muschenbroeck of Leyden. This was an important invention, as it was the forerunner of our own Franklin's discoveries and a necessary part of his outfit with which he established the identity of lightning and electricity. Every American schoolboy has heard, from Fourth of July orations, how "Franklin caught the forked lightning from the clouds and tamed it and made it subservient to the will of man." How my boyish soul used to be stirred to its depths by this oratorical display of electrical fireworks!

Franklin had long entertained the idea that the lightning of the clouds was identical with what is called frictional electricity, and he waited long for a church spire to be erected in his adopted home, the Quaker City, in order that he might make the test and settle the question. But the Quakers did not believe in spires, and Franklin's patience had a limit.

Franklin had the theory that most investigators had at that time, that electricity was a fluid and that certain substances had the power to hold it.

There were two theories prevalent in those days--both fluid theories. One theory was that there were two fluids, a positive and a negative. Franklin held to the theory of a single fluid, and that the phenomenon of electricity was present only when the balance or natural amount of electricity was disturbed. According to this theory, a body charged with positive electricity had an excessive amount, and, of course, some other body somewhere else had less than nature had allotted to it; hence it was charged with negative electricity. A Leyden jar, for instance, having one of its coatings (say the inside) charged with positive or + electricity, the other coating will be charged with negative or - electricity. The former was only a name for an amount above normal and the latter a name for a shortage or lack of the normal amount.

As we have said, Franklin believed in the identity of lightning and electricity, and he waited long for an opportunity to demonstrate his theory. He had the Leyden jar, and now all he needed was to establish some suitable connection between a thunder-cloud and the earth.

Previous to 1750 Franklin had written a paper in which he showed the likeness between the lightning spark and that of frictional electricity. He showed that both sparks move in crooked lines--as we see it in a storm-cloud, that both strike the highest or nearest points, that both inflame combustibles, fuse metals, render needles magnetic and destroy animal life. All this did not definitely establish their identity in the mind of Franklin, and he waited long for an opportunity, and finally, finding that no one presented itself, he did what many men have had to do in other matters; he made one.

In the month of June, 1752, tired of waiting for a steeple to be erected, Franklin devised a plan that was much better and probably saved the experiment from failure; for the steeple would probably not have been high enough. He constructed a kite by making a cross of light cedar rods, fastening the four ends to the four corners of a large silk handkerchief. He fixed a loop to tie the kite string to and balanced it with a tail, as boys do nowadays. He fixed a pointed wire to the upper end of one of the cross sticks for a lightning-rod, and then waited for a thunder-storm. When it came, with the help of his boy, he sent up the kite. He tied a loop of silk ribbon on the end of the string next his hand--as silk was known to be an insulator or non-conductor--and having tied a key to the string he waited the result, standing within a door to prevent the silk loop from getting wet and thus destroying its insulating qualities. The cloud had nearly passed and he feared his long waited for experiment had failed, when he noticed the loose fibers of the string standing out in every direction, and saw that they were attracted by the approach of his finger. The rain now wet the string and made a better conductor of it. Soon he could draw sparks with his knuckle from the key. He charged a Leyden jar with this electrical current from the thunder-cloud, and performed all the experiments with it that he had done with ordinary electricity, thus establishing the identity of the two and confirming beyond a doubt what he had long before believed was true. In after experiments Franklin found that sometimes the electricity of the clouds was positive and at other times negative. From this experiment Franklin conceived the idea of erecting lightning-rods to protect buildings, which are used to this day.

The news spread all over Europe, not through the medium of electricity, however, but as soon as sailing vessels and stage-coaches could carry it. Many philosophers repeated the experiments and at least one man sacrificed his life through his interest in the new discovery. In 1753 Professor Richman of St. Petersburg erected on his house a metal rod which terminated in a Leyden jar in one of the rooms. On the 31st of May he was attending a meeting of the Academy of Sciences. He heard a roll of thunder and hurried home to watch his apparatus. He and one of the assistants were watching the apparatus when a stroke of lightning came down the rod and leaped to the professor's head. He was standing too near it and was instantly killed.

Passing over many names of men who followed in the wake of Franklin we come to the next era-making discovery, namely, that of galvanic electricity. In the year 1790 an incident occurred in the household of one Luigi Galvani, an Italian physician and anatomist, that led to a new and important branch of electrical science. Galvani's wife was preparing some frogs for soup, and having skinned them placed them on a table near a newly charged electric machine. A scalpel was on the table and had been in contact with the machine. She accidentally touched one of the frogs to the point of the scalpel, when, lo! the frog kicked, and the kick of that dead frog changed the whole face of electrical science. She called her husband and he repeated the experiment, and also appropriated the discovery as well, and he has had the credit of it ever since, when really his wife made the discovery. Galvani supposed it to be animal electricity and clung to that theory the rest of his life, making many experiments and publishing their results; but the discovery led others to solve the problem.

Alessandro Volta, a professor of natural philosophy at Pavia, Italy, was, it must be said, the founder of the science of galvanic or voltaic electricity. Stimulated by the discovery of Galvani he attributed the action of the frog's muscles, not to animal electricity, but to some chemical action between the metals that touched it. To prove his theory, he constructed a pile made of alternate layers of zinc, copper, and a cloth or pasteboard saturated in some saline solution. By repeating these trios--copper, zinc, and the saturated cloth--he attained a pile that would give a powerful shock. It is called the Voltaic Pile.

I have a clear idea of the construction of this form of pile, founded on experience. It was my habit when a boy to make everything that I found described, if it were possible. The bottom of my mother's wash-boiler was copper, and just the thing to make the square plates of copper to match the zinc ones, made from another piece of domestic furniture used under the stove. I shocked my mother twice--first with the voltaic pile that I had constructed, and again when she found out where the metal plates came from. The sequel to all this was--but why dwell upon a painful subject!

Galvanism and voltaic electricity are the same. Volta was the first to construct what is termed the galvanic battery. The unit of electrical pressure or electromotive force is called the volt, and takes its name from Volta, the great founder of the science of galvanic or voltaic electricity. From this pile constructed by Volta innumerable forms of batteries have been devised. The evolution of the galvanic battery in all its forms, from Volta to the present day, would fill a large volume if all were described.

The discoveries of Michael Faraday (1791-1867), the distinguished English chemist and physicist, led to another phase of the science that has revolutionized modern life. Faraday made an experiment that contains the germ of all forms of the modern dynamo, which is a machine of comparatively recent development. He found that by winding a piece of insulated wire around a piece of soft iron and bringing the two ends (of the wire) very close together, and then placing the iron across the poles of a permanent magnet and suddenly jerking it away, a spark would pass between the two ends of the wire that was wound around the piece of soft iron. Here was an incipient dynamo-electric machine--the germ of that which plays such an important part in our modern civilization.

Having brought our history down to the present day, it would seem scarcely necessary to recite that which everybody knows. It is well, however, to call a halt once in a while and compare our present conditions of civilization with those of the past. Our world is filled with croakers who are always sighing for the good old days. But we can easily imagine that if they could go back to those days their croaking would be still louder than it is.

Before the advent of electricity many things were impossible that are easy now. In the old days the world was very, very large; now, thanks to electricity, it is knocking at the door of every man's house. The lumbering stage-coach that was formerly our limited express--limited to thirty or forty miles a day-- has been supplanted by one that covers 1000 miles in the same time, and this high rate of speed is made possible only by the use of the electric telegraph.

In the old days all Europe could be involved in a great war and the news of it would be weeks in reaching our shores, but now the firing of the first gun is heard at every fireside the world over, almost before the smoke has cleared away. Our planet is threaded with iron nerves that run over mountains and under seas, whose trembling atoms, thrilled with the electric fire, speak to us daily and hourly of the great throbbing life of the whole civilized world.

Electricity has given us a voice that can be heard a thousand miles, and not only heard, but recognized. It has given us a pen that will write our autograph in New York, although we are still in Chicago. It has given us the best light, both from an optical and a sanitary standpoint, that the world has ever seen. The old-fashioned, jogging horse-car has been supplanted by the electric "trolley," and we no longer have our feelings harrowed with pity for the poor old steeds that pulled those lumbering coaches through the streets, with men and women crowded in and hanging on to straps, while everybody trod on every other body's toes.

"In olden times we took a car Drawn by a horse, if going far, And felt that we were blest; Now the conductor takes the fare And puts a broomstick in the air-- And lightning does the rest.

"In other days, along the street, A glimmering lantern led the feet, When on a midnight stroll; But now we catch, when night is nigh, A piece of lightning from the sky And stick it on a pole.

"Time was when one must hold his ear Close to a whispering voice to hear, Like deaf men--nigh and nigher; But now from town to town he talks And puts his nose into a box And whispers through a wire."

So jogs the old world along. We sometimes think it is slow, but when we look back a few years and see what has been accomplished it seems to have had a marvelously rapid development.

Something like fifty years ago a professor of physics in one of our colleges was giving his class a course in electricity. The electric telegraph was too little known at that time to cut much of a figure in the classroom, so the stock experiments were those made with the frictional electric machine and the Leyden jar. One day the professor had, in one hour's time, taken his class through a course of electricity, and at the end he said: "Gentlemen, you were born too late to witness the development of this great science." I often wonder if the good professor is ever allowed to part the veil that separates us from the great beyond and to look down upon this busy world of ours in which electricity plays such an important part in our every-day life; and if so, what he thinks of that little speech he made to the boys fifty years or more ago.

If we make an analysis of the history of the science of electricity we shall see that it has progressed in successive eras, shortening as they approach our time. For a period of 2300 years, from Thales to Franklin, but little or no progress was made beyond the further development of the phenomena of frictional electricity--the most important invention being that of the Leyden jar. From Franklin to Volta was forty-eight years, and from Volta to Faraday about thirty-two years. From this time on the development was very rapid as compared with the old days. Soon after Faraday, Morse, Henry, Wheatstone, and others began experiments that have grown, during fifty or sixty years, into a most colossal system of electric telegraphs, telephones, electric lights and electric railroads. In the latter days marvel has succeeded marvel with such rapid strides that the ink is scarcely dry from the description of one before another crowds itself upon our attention. Where it will all end no one knows, but that it has ended no one believes. The human mind has become so accustomed to these periodic revelations of the marvelous that it must have the stimulus once in a while or it suffers as the toper does when deprived of his cups. The commercial instinct of the news-vender is not slow to see the situation, and if the development is too slow to suit the public demand his fertile brain supplies the lack. So that every few days we hear of some great discovery made by some one it may be unknown to fame. It has served its purpose. The public mind has had its mental toddy and has been saved from a fit of intellectual delirium tremens that it was in danger of from lack of its accustomed stimulus.

Having given you a very limited outline of the history of electricity, from ancient times down to the present, we will endeavor now to give you an elementary notion of the science as it stands to-day. To the common mind the science is a blank page. So little is known of it by the ordinary reader, who is fairly intelligent in other matters, that to account for anything that we do not understand it is only necessary to say that it is an electrical phenomenon and he accepts it. Electricity is a synonym for all that we cannot understand. Inasmuch as magnetism is so closely related to electricity in its uses as related to every-day life, we will carry the two subjects along together, as the one will to a large extent help to explain the other. In our next chapter we will look at the history of magnetism.

CHAPTER III.

HISTORY OF MAGNETISM.

It is said that the word magnetism is derived from the name of a Greek shepherd, called Magnes, who once observed on Mount Ida the attractive properties of loadstone when applied to his iron shepherd's crook. It is more likely that the name came from Magnesia, a country in Lydia, where it was first discovered. It was also called Lapis Heracleus. Heraclea was the capital of Magnesia. Loadstone is a magnetic ore or oxide of iron found in the natural state, and has at some time by natural processes been rendered magnetic-- that is, given the power of attracting iron, and, when suspended, of pointing to the North and South Poles. The power of the natural magnet was known at a very early age in the history of man. It was referred to by Homer, Pythagoras, and Aristotle. Pliny also speaks of it, and refers to one Dinocares, who recommended to Ptolemy Philadelphus to build a temple at Alexandria and suspend in its vault a statue of the queen by the attractive power of "loadstones." There is also mention of a statue being suspended in like manner in the temple of Serapis, Alexandria.

It is claimed that the Chinese knew of and used the magnetic needle in the earliest times and that travelers by land employed this needle suspended by a string to guide them in their journeys across the country a thousand years before Christ. Notwithstanding the claims of the Chinese and Arabians to the discovery of the use of the magnetic needle, modern authors question whether the ancients were familiar with any artificial construction of a magnetic needle, however much they may have studied and used the loadstones. No doubt the loadstone in its natural state was used by mariners to steer their ships by, long before its artificial counterpart was invented. In a history of the discovery of Iceland, by Are Frode, who was born in 1068, it is stated that a mariner by name of Folke Gadenhalen sailed from Norway in search of Iceland in the year 868, and that he carried with him three ravens as guides, for he says, "in those times seamen had no loadstones in the northern countries." The magnetic needle as applied to the mariner's compass was known in the eleventh century, as proved by various authors. In an old French poem, the manuscript of which still exists, the mariner's compass is clearly mentioned. The author was Guyot, of Provence, who was alive in 1181.

Like electricity, magnetism has had a long history, but little use was made of it till modern times beyond that of the mariner's compass. It can readily be seen what an important factor it was in the science of navigation. Long after the discovery of the compass needle there were many perplexing problems arising, and all sorts of theories were advanced to account for the various phenomena. The variation of the needle was one of these problems. It is said that Columbus was the first to discover the variation of the needle, as well as America. This is disputed, however, as every man's pretensions usually are. However this may be, Columbus had to invent some plausible theory to account for this variation to prevent a mutiny among his crew. They were very superstitious and thought that they were sailing into a new world where the laws of nature were different from those of Spain. One phenomenon that disturbed Columbus was the dip of the needle. As we move in a northerly direction a magnetic needle dips, and it was the observation of this phenomenon in different latitudes that finally resulted in the invention of the dipping needle. It is well known that one pole of a magnetic needle points to the north and the other to the south. In other words, what is called the north pole of a needle points to one of the magnetic poles of the earth which is in the direction of the north pole, though not the same as the geographical pole. A dipping needle revolves on an axis so that it can point to any declination. If we should construct one that is perfectly balanced, so as to lie in a perfectly horizontal direction before it is magnetized, it will dip--in this latitude-- downward toward the north after magnetization. If we keep moving northward it will continue to dip downward till we come to the true magnetic pole, when what is called the north pole of the needle will point directly downward. If we go back to the equator the needle will lie horizontally again. We call the end of the needle that points to the north the north pole. It is really the south pole, because unlike poles attract each other. If the magnetic poles of the earth are at the north and south geographical poles, the south pole of the needle will point north. But it is less confusing to call the end of the needle that points north the north pole. The nomenclature is purely arbitrary.

It was not until it was learned that magnets could be made by electricity that they became commercially important outside of their use in navigation. The advent of electricity has brought magnetism to the front as one of the great factors in our modern civilization. And we might say with equal force that the discovery of magnetism has brought electricity to the front. The truth is that they depend upon each other. Electricity would be robbed of a large part of its importance as a factor in modern life if it were not for its relation to magnetism. Even electric lighting would be impossible, commercially, if it were not for the part magnetism plays in the production of electricity for this purpose. It could not be successfully carried on with any battery but the storage-battery, and the storage-battery is dependent upon the dynamo, and the dynamo is a magneto-electric machine. When we come to analyze the relation between magnetism and electricity we cannot separate them without robbing each of a large part of its usefulness. They are interdependent forces.

As in the case of electricity there have been many theories regarding magnetism. One philosopher in the old days accounts for the variation of the compass-needle on the theory that there are two globes, one revolving within the other, and that any derangement of their normal movements in relation to each other affects the needle. Evidently there were cranks in those days as well as now. Another theory of magnetism was that there were two fluids--a boreal and an austral--one developing north polarity and the other south polarity. In the next chapter the nature of magnetism in the light of modern investigation will be discussed.

CHAPTER IV.

THEORY AND NATURE OF MAGNETISM.

Iron and steel have a peculiar property called magnetism. It is an attraction in many ways unlike the attraction of cohesion or the attraction of gravitation. It is very certain that magnetism is an inherent property of the molecules of iron and steel, and, to a small degree, other forms of matter. That is to say, the molecules are little natural magnets of themselves. It is as unnecessary to inquire why they are magnets as it is to inquire why the molecules of all ordinary substances possess the attraction of cohesion. The one is as easy to explain as the other. People of all ages have insisted upon making a greater mystery of all electrical and magnetic phenomena than they do of other natural forces. Ampere's theory is that electric currents are flowing around the molecules which render them magnetic; but it is just as easy to suppose that magnetism is an inherent quality of the molecule. (The word molecule is here used as referring to the smallest particle of iron.)

These little molecular magnets, so small that 100,000 million million million of them can be put into a cubic inch of space, have their attractions satisfied by forming into little molecular rings, with their unlike poles together, so that when the iron is in a natural or unmagnetized condition it does not attract other iron. If I should take a ring of hardened steel and cut it into two or more pieces and magnetize them, each one of the pieces would be an independent magnet. If now I put them together in the form of a ring they will cling together by their mutual attraction for each other. Before I put them together into a ring each piece would attract and adhere to other pieces of iron or steel. But as soon as they are put together in the ring they are satisfied with their own mutual attraction, and the ring as a whole will not attract other pieces of iron.

Suppose the pieces forming the ring--it may be only two, if you choose--are as small as the molecules we have described, the same thing would be true of them. Each molecular ring would have its magnetic attractions satisfied and would not attract other molecules outside of its own little circle. When the iron is in the neutral state it will not as a mass attract another piece of iron, because the millions of little natural magnets of which it is made up have their attractive force all turned in upon themselves.

Now, if we make a helix, or coil, of insulated wire and put a piece of iron into it, and pass a current of electricity through the helix, the iron becomes a magnet. Why? Because the electric current has the power to break up these molecular magnetic rings and turn all their like poles in one direction, so that their attractions are no longer satisfied among themselves, and with a combined effort they reach outside and attract any piece of iron that is within reach. In this state we say it is magnetized. Most people think that we have put something into the iron, but we have not; we have only developed and made active its inherent power. It must be kept in mind that it takes power to develop this magnetic power from its state of neutrality and that something is never made from nothing. When this power is developed it will do work in falling back to its natural state. The power is natural to the molecules of the metal. It is only being exerted in a new direction. The millions of little natural magnets have been forced to combine their attractions into one whole and exert it on something outside of themselves. They are under a strain in this condition, like a bent bow, and there is a tendency to fly back to the natural position, and if it is soft iron and not steel, they will fly back as soon as the power that wrenched them apart and is holding them apart is taken away. This power is the electric current. Now break the current, and the little natural magnets, that have been so ruthlessly torn from their home circle attachments, fly back to them again with the speed of lightning, and the iron rod as a whole is no longer a magnet. The power to become so under the electrical strain is in it still--only latent.

The kind of magnet that we have been describing is called an electromagnet. It is a magnet only so long as the electric current is passing around it. There is another kind of magnet called a permanent magnet that will remain a magnet after the current is taken away. The permanent magnet is made of steel and hardened; then its poles are placed, to the poles of a powerful magnet, either electro or permanent, when its molecular rings are wrenched apart and arranged in a polarized position as heretofore described. Now take it away from the magnet and it will be found to retain its magnetism. The molecules tend to fly back the same as those of the soft iron, but they cannot because hardened steel is so much finer grained than soft iron, and the molecules are so close together that they are held in position by a friction that is called its coercive force. The soft iron is comparatively free from this coercive force, because its molecules are free to move on each other, so that when they are wrenched out of their natural position they fly back by their own attractions as soon as the force holding them apart is taken away. The molecules of hardened steel are unable to fly back, although they tend to do it just as much as in the iron, and so it is called a permanent magnet. Its molecules also are under a strain, like a bent bow. (The form of such a magnet is usually that of a horse-shoe, or U.)

Let us use a homely illustration that may help us to understand. Let ten boys represent the molecules in a piece of iron. Let them pair off into five pairs and each one clasp his mate in his arms; each one, say, is exerting a force of ten pounds, and it would require a force of twenty pounds to pull any one of the pairs apart. The five pairs are exerting a force of one hundred pounds, but this force is not felt outside of themselves. Now let them unclasp themselves and take hold of a rope that is tied to a post, and all pull with the same force that they were using, to wit, ten pounds each, and all pull in the same direction, and they would put a strain of one hundred pounds upon the post, the same power that they were exerting upon themselves before they combined their efforts on something outside of themselves. So with the magnet. So long as the force of each molecule is wholly spent upon its neighbor there is nothing left for exterior use. But as soon as they all line up and pull conjointly in the same direction their combined force is felt outside. The analogy may not be perfect, but it will help you to get a mental picture of what takes place in iron when it is magnetized.

We have now described the magnet and the inherent power residing in the molecular structure of iron. It is this magic power slumbering in its molecules and the ability of the electric current to arouse them to action at will and to hold them in action and at will let them fly back to their normal position, that gives to electricity and magnetism--twin sisters in nature's household--their great value as the servants of man. There would be no virtue in winding up a weight if it could not run down and do work in its fall. Simply bending a bow would never send the arrow flying over its course; it must be released as well. The magnet could not accomplish the great work it does if we could only charge it and not have the ability to discharge it. Without this ability the electric motor would not revolve, the electric light would not burn, the click of the telegraph would not be heard, the telephone would not talk, nor would the telautograph write.

I have said that the permanent magnet would hold its charge after once having been magnetized. This is true only in a sense and under favorable conditions. If made of the best of steel for the purpose and hardened and tempered in just the right way, it will hold its charge if it is given something to do. If a piece of iron is placed across its poles it also becomes a magnet and its molecules turn and work in harmony with those of the mother magnet. These magnetic lines of force reach around in a circuit. Even before the iron, or "keeper," as it is called, is put across its poles there are lines of force reaching around through the air or ether from one pole to another. (For a description of Ether see Chap. V.) This is called the "field" of the magnet, and when the iron is placed in this field the lines of force pass through it in a closed circuit, and if the "keeper" is large enough to take care of all the lines of force in the field the magnet will not attract other bodies, because its attraction is satisfied, like its prototype in the molecular ring described above.

We speak of lines of force, not that force is necessarily exerted in a bundle of lines but as a convenient way of telling the strength of a magnetic field. The practical limit of the magnetization of soft iron (called saturation) is 18,000 lines to the square centimeter. As long as we give our magnet something to do, up to the measure of its capacity, it will keep up its power. We may make other magnets with it, thousands, yea, millions of them, and it not only does not lose its power but may be even stronger for having done this work. If, however, we hang it up without its "keeper," and give it nothing to do, it gradually returns to its natural condition in the home circle of molecular rings. Little by little the coercive force is overcome by the constant tendency of the molecule to go back to its natural position among its fellows.

The magnet furnishes many beautiful lessons, as indeed do all the natural phenomena. Every man has within him a latent power that needs only to be aroused and directed in the right way to make his influence felt upon his fellows. Like the magnet, the man who uses his power to help his fellows up to the measure of his limitations not only has been a benefactor to his race, but is himself a stronger and better man for having done so. But, again, like the magnet, if he allows these God-given powers to lie still and rust for want of legitimate use he gradually loses the power he had and becomes simply a moving thing without influence or use in a world in which he vegetates. But let us leave philosophy and go back to science.

One of the striking exhibitions of magnetism is found in the earth. The earth itself is a great magnet; and there is good reason for believing that it is an electromagnet of great power. The magnetic poles of the earth are not exactly coincident with the geographical poles, and they are not constant. There is a gradual deviation going on, but as it follows a certain law mariners are able to tell just what the deviation should be at a certain time. The magnetic pole revolves around the polar axis of the earth once in about 320 years. A thermal current (one produced by heat) of electricity seems to flow around the earth caused by the irregularities of temperature at the earth's surface, as the sun makes his daily round. These earth currents vary at times, and other phenomena are the occasion. This will be discussed when we come to electric storms.

The value of the earth's magnetism is seen most in the science of navigation. A magnetic needle is only a slender permanent magnet suspended very delicately, and when not under local influence it points north and south on the magnetic axis. The law of its action may be explained as follows: Take a straight bar magnet of fairly good power and suspend a magnetic needle over it. The needle will arrange itself parallel to the bar magnet. The north pole of the needle will point toward the south pole of the bar magnet. In the presence of the magnet the needle is not affected by the earth, but yields to a superior force. If, however, the bar magnet is taken out of the way of the needle it will immediately arrange itself north and south. Of course if the earth's magnetic axis changes the needle will vary with it. This variation is uniform and in navigation is reduced to a science, so that the mariner knows how much to allow for the variation. Columbus, as heretofore mentioned, was supposed to have first noticed this variation and it made him trouble. He did not know how to account for it, and as his crew thought the laws of nature were changing because they were so far from home he saw the necessity for some sort of explanation. So, like the brave man that he was, he hatched up a theory that satisfied the crew, and although in the light of the closing years of the nineteenth century it was a questionable one, it worked well enough in practice to serve his purpose.

We have already stated that the earth was a great magnet, and that probably it was an electromagnet, caused by earth currents circulating around the globe. You want to know how the earth can be a magnet unless it has an iron core like an electromagnet. Magnetism or magnetic lines of force may be developed without the presence of iron. When we pass a current of electricity through a wire, magnetic lines of force are thrown out at right angles with the direction of the current. This will be fully explained further on. If we wind the wire into a coil, or helix, these magnetic lines are concentrated. If now we suspend this helix, or, better, float it on water so that it can move freely, and pass a current of electricity through it, the helix will arrange itself north and south the same as a magnetic needle. Its attractive properties are feeble in comparison with that of the iron, but it obeys the laws of a magnet. The earth is probably a magnet of this kind, consisting mostly of lines of force.

However, the iron in the earth is affected magnetically, as we have evidence in the loadstone. The earth has the power also to magnetize iron through the medium of its magnetic field, that reaches out in lines of force from pole to pole like those of the artificial magnet. If we hold a bar of iron in line with the magnetic axis of the earth and dip it in line with the dipping needle and then strike it a few blows on the end, it will be found to be feebly magnetic. The blows have partly loosened the molecules and during the moment that they unclasped themselves the earth's magnetism has through its lines of force caught them for a time and held them a little out of their natural position--as they are in a state of rest. The peculiar changing light that we sometimes see in the northern sky, that is called the Aurora Borealis (Northern Light), is indirectly due to intense magnetic lines of force that radiate from the north magnetic pole of the earth. Those lines of force are able to cause the rarified air molecules to become feebly incandescent, giving them the appearance that we see in a tube that is a partial vacuum when electricity is passed through it. While these auroral displays may be seen almost any night in the far north, they vary greatly in their intensity, so it is only once in a while that they are visible in the temperate latitudes.

What are called magnetic storms occur occasionally, and at such times the telegraph service will sometimes be paralyzed on all the east and west lines for many hours. Strong earth-currents will flow east and west, and be so powerful and so erratic that it is sometimes impossible to use the telegraph. It sometimes happens that the operators can throw off their batteries and work on the earth-current alone. Sometimes it is necessary to make a complete metallic circuit to get away from the influence of the earth in order to use the telegraph. Currents equal to the force of 2,000 cells of ordinary battery have been developed sometimes in telegraph wires. This of course is a mere fraction of what is passing through the earth under the wire through which the current flowed. On the 17th and 18th of November, 1882, a magnetic storm occurred that extended around the globe, as it was felt wherever there were telegraph wires. These magnetic storms are attended by brilliant displays of the aurora, and this fact strengthens the theory that the earth is a great electromagnet; for the stronger the electrical current the more powerful we should expect the magnetism to be, and this is shown by the action of the magnetic needle at such times. The stronger the magnet the more intense will be the lines of force, and naturally the more intense the light, if indeed these lines of force are the cause of the light. There is evidently some close relation between the two.

Another coincidence is that at the times of these storms there is an unusual display of sun-spots. These sun-spots seem to be great holes that have been blown through the photosphere of the sun. The photosphere is a great luminous body of gaseous matter that is believed to envelop the sun, so that we do not see the core of the sun unless it is when we look into one of these spots. In some way, evidently, the sun affects the earth by radiating magnetic lines of force which are cut by the earth's revolution, and so creating currents of electricity. The sun is the field-magnet, and the earth is the revolving armature of nature's great dynamo-electric machine. It would seem that the radiant energy that comes out through these spots or these holes in the sun's envelope, are more potent to develop earth-currents than the ordinary rays; and so, when for a brief while in the revolution of the earth about the sun, these extra potent rays strike the earth, an unusual energy is developed, and these unusual phenomena are the consequence. These phenomena seem to occur periodically; some years (about eleven) intervening.

All the forces and phenomena of nature are thus seen to be in a state of unrest. And it is to this unrest, which does not stop with visible things, but pervades even the atoms of matter throughout the universe, that we are indebted for the ability to carry on all the activities of life, and for life itself. For universal quiet would mean universal death. The cyclone and tornado that devastate and strike terror to a whole region are only eccentricities of nature when she is setting her house to rights. The play of natural forces has disturbed her equilibrium, and she is but making an effort to restore it.

CHAPTER V.

THEORY OF ELECTRICITY.

In the series of chapters on Heat (Vol. II) and in the chapter on Magnetism the word molecule was frequently used synonymously with atom. In chemistry a distinction is made, and as we can better explain the theory, at least, of electricity by keeping this distinction in mind we will refer to it here.

It has been stated that there are between sixty and seventy elementary substances. An elementary substance cannot be destroyed as such. It can be united with other elements and form chemical compounds of almost endless variety. The smallest particle of an elementary substance is called in chemistry an atom. The smallest particle of a compound substance is called a molecule. The atom is the unit of the element, and the molecule is the unit of the compound as such. It follows, then, that there are as many different kinds of atoms as there are elements, and as many different kinds of molecules as there are compounds. If the elements have a molecular Structure then two or more atoms of the same kind must combine to make a molecule of an elementary substance. Two atoms of hydrogen combine with one of oxygen to form one molecule of water. It cannot exist as water in any smaller quantity. If we subdivide it, it no longer exists as water, but as the original gases from which it was compounded.

We have shown in the series on Sound, Heat and Light that they are all modes of motion. Sound is transmitted in longitudinal waves through air and other material substance as vibration. Heat is a motion of the ultimate particles or atoms of matter, and Light is a motion of the luminiferous ether transmitted in waves that are transverse. Electricity is also undoubtedly a mode of motion related in a peculiar way to the atoms of the conductor.

Notice that there is a difference between conduction and radiation. The former transmits energy by a transference of motion from atom to atom or molecule to molecule within the body, while the latter does it by a vibration of the ether outside--as light, radiant heat, and electromagnetic lines of force.

For the benefit of those persons who have not read Vol. II, where the nature of ether is discussed somewhat, let us refer to it here, as it plays an important part in the explanation of electrical phenomena. Ether is a tenuous and highly elastic substance that fills all interstellar and interatomic space. It has few of the qualities of ordinary matter. It is continuous and has no molecular structure. It offers no perceptible resistance, and the closest-grained substances of ordinary matter are more open to the ether than a coarse sieve is to the finest flour. It fills all space, and, like eternity, it has no limits. Some physicists suppose--and there is much plausibility in the supposition--that the ether is the one substance out of which all forms of matter come. That the atoms of matter are vortices or little whirlpools in the ether; and that rigidity and other qualities of matter all arise in the ether from different degrees or kinds of motion.

Electricity is not a fluid, or any form of material substance, but a form of energy. Energy is expressed in different ways, and, while as energy it is one and the same, we call it by different names--as heat energy, chemical energy, electrical energy, and so on. They will all do work, and in that respect are alike. One difficulty in explaining electrical phenomena is the nomenclature that the science is loaded down with. All the old names were adopted when electricity was regarded as a fluid, hence the word "current." It is spoken of as "flowing" when it does not flow any more than light flows.

If a man wants to write a treatise on electricity--outside of the mere phenomena and applications--and wants to make a large book of it, he would better tell what he does not know about it, for in that way he can make a volume of almost any size. But if he wants to tell what it really is, and what he really knows it is, a primer will be large enough. This much we know--that it is one of many expressions of energy.

Chemistry teaches that heat is directly related to the atoms of matter. Atoms of different substances differ greatly in weight. For instance, the hydrogen atom is the unit of atomic weight, because it is the lightest of all of them. Taking the hydrogen atom as the unit, in round numbers the iron atom weighs as much as 56 atoms of hydrogen, copper a little over 63, silver 108, gold 197. Heat acts upon matter according to the number of atoms in a given space, and not as its weight. Knowing the relative weights of the atoms of the different metals named, it would be possible to determine by weight the dimensions of different pieces of metal so that they will contain an equal number of atoms. If we take pieces of iron, copper, silver and gold, each of such weight as that all the pieces will contain the same number of atoms, and subject them to heat till all are raised to the same temperature, it will be found that they have all absorbed practically the same quantity of heat without regard to the different weights of matter. It will be observed that the piece of silver, for instance, will have to weigh nearly twice as much as the iron in order to contain the same number of atoms, but it will absorb the same amount of heat as the piece of iron containing the same number of atoms, if both are raised to the same temperature. In view of the above fact it seems that heat acts especially upon the atoms of matter and is a peculiar form of atomic motion. Heat is one kind of motion of the atoms, while electricity may be another form of motion of the same. The two motions may be carried on together. The earth has a compound motion. It revolves upon its axis once in twenty-four hours, and it also revolves around the sun once each year. So you see that there are different kinds of motion that may be communicated to the same body--all producing different results.

The motion of the individual atom as heat may be, and is, as rapid as light itself when the temperature is sufficiently high, but it does not travel along a conductor rapidly as the electro-atomic motion will. If we apply heat to the end of a metal rod it will travel slowly along the rod. But if we make the rod a conductor of electricity it travels from atom to atom with a speed nearer that of the light ray through the ether. Some modern writers have attempted to explain all the phenomena of electricity as having their origin in a certain play of forces upon the ether, and there is no doubt but that the ether plays an important part in all electrical phenomena as a medium through which energy is transferred; but ether-waves that are set in motion by the electrical excitation of ordinary matter are no more electricity than the ether-waves set up by the sun in the cold regions of space are heat. They become heat only when they strike matter. Heat, as such, begins and ends in matter;--so (I believe) does electricity.

Do not be discouraged with these feeble attempts to explain the theory of electricity. All I even hope to do is to establish in your minds this fundamental thought, to wit, that there is really but one Energy, and that it is always expressed by some form of motion or the ability to create motion. Motions differ, and hence are called by different names.

If I should set an emery-wheel to revolving and hold a piece of steel against it the piece of steel would become heated and incandescent particles would fly off, making a brilliant display of fireworks. The heat that has been developed is the measure of the mechanical energy that I have used against the emery-wheel. Now, let us substitute for the emery-wheel another wheel of the same size made of vulcanized rubber, glass or resin. I set it to revolving at the same speed, and instead of the piece of steel, I now hold a silk handkerchief or a catskin against the wheel with the same force that I did the steel. If now I provide a Leyden jar and some points to gather up the electricity that will be produced (instead of the heat generated in the other case), it would be found that the energy developed in the one case would exactly balance that of the other, if it were all gathered up and put into work. The electricity stored in the jar is in a state of strain, like a bent bow, and will recoil, when it has a chance, with a power commensurate with the time it has been storing and the amount of energy used in pressing against the wheel.

If now I connect my two hands, one with the inside and the other with the outside of the jar, this stored energy will strike me with a force equal to all the energy I have previously expended in pressing against the wheel, minus the loss in heat. If I did it for a long enough time this electrical spring would be wound up to such a tension that the recoil would destroy life if one put himself in the path of its discharge. If all the heat in the first case were gathered up and made to bend a stiff spring, and one should put himself in its way when released, this mechanical spring would strike with the same power that the electrical spring did when the Leyden jar was discharged. This statement assumes that all the energy in the second experiment was stored as electricity in the jar. You will be able to see from the above illustration that heat, electrical energy, and mechanical energy are really the same. Then you ask, how do they differ? Simply in their phenomena--their outward manifestations.

While there is much that we cannot know about any of the phenomena of nature, it is a great step in advance if we can establish a close relationship between them. It helps to free electricity from many vagaries that exist in the minds of most people regarding it; vagaries that in ignorant minds amount to superstition. While it possesses wonderful powers, they give it attributes that it does not possess. Not long ago a favorite headline of the medical electrician's advertisement was "Electricity Is Life," and it was a common thing to see street-venders dealing out this "life" in shocking quantities to the innocent multitudes--ten cents' worth in as many seconds.

Science divides electricity into two kinds--static and dynamic. Static comes from a Greek word, meaning to stand, and refers to electricity as a stationary charge. Dynamic is from the Greek word meaning power, and refers to electricity in motion. When Franklin made his celebrated kite experiment, the electricity came down the string, and from the key on the end of the string he stored it in a Leyden jar. While the electricity was moving down the string it was dynamic, but as soon as it was stored in the Leyden jar it became static. Current electricity is dynamic. A closed telegraphic circuit is charged dynamically, while the prime conductor of a frictional electric machine is charged statically. The distinction is arbitrary and in a sense a misnomer. When we rub a piece of hard rubber with a catskin it is statically charged because the substances are what are called non-conductors, and the charge cannot be conducted readily away. All substances are divided into two classes, to wit, conductors or non-electrics, and non-conductors or electrics, more commonly called dielectrics. These, however, are relative terms, as no substance is either a perfect conductor or a perfect non-conductor.

The metals, beginning with silver as the best, are conductors. Ebonite, paraffine, shellac, etc., are insulators, or very poor conductors. The best conductors offer some resistance to the passage of the current and the best insulators conduct to some extent. If we make a comparison of electric conductors we find that the metals that conduct heat best also conduct electricity best. This, it seems to me, is a confirmation of the atomic theory of electricity so far as it means anything. If a good conductor, as silver, is subjected to intense cold by putting it into liquid air, its conductivity is greatly increased. It is well known that heating a conductor ordinarily diminishes its power to conduct electricity. This shows that, in order that electrical motion of the atom may have free play, the heat motion must be suppressed.

CHAPTER VI.

ELECTRIC CURRENTS.

The simplest form of an electric machine is one in which the operator is a prominent part of the operation. Electricity, like magnetism, operates in a closed circuit, even when it is static--so-called. Take a stick of sealing-wax, say, in your left hand, and rub it with a piece of fur or silk with your right hand, and you have the simplest form of electric machine--the one that was known to the ancients, and the one from which the science, great as it is to-day, had its beginnings. The stick of sealing-wax is one element of the battery, and the piece of fur or silk is the other, while your hands, arm and body form the conductor that connects the two poles, and the friction is the exciting agent and may be said to take the place of the fluid of a battery. The electrical conditions are not wholly static, as a slow current is passing around through your arms and body from one pole to the other. Even if the conditions were wholly static there would be polarized lines of force, in a state of strain, reaching around in a closed circuit.

If we rub the wax with the fur and then take it away the wax has a charge of electricity and will attract light objects. If we had rubbed a piece of metal or some good conductor it would have been warmed instead of electrified. In both cases the particles of the substances have been affected, and if the atomic theory is correct--and it seems plausible--in the former case the atoms are partly put into electrical motion and partly into a state of electrical strain that we call static (standing) electricity; while in the latter case the atoms are put into the peculiar motion that belongs to heat. The former we call electricity, and the latter we call heat. The electro-atomic motion under some circumstances readily turns to heat, which seems to be the tendency of all forms of energy. The electric light is a result of this tendency. All non-conductors, or electrics, have a complex molecular structure, and, while their atoms when subjected to friction are put into a state of electrostatic strain, they are not able readily to respond as a conductor of dynamic electricity. The electric-light filament in the incandescent lamp is a much poorer conductor than the copper wire that leads up to it. The copper wire is readily responsive to the electrical influence, but the carbon filament is not. So electrical action that freely passes along the wire, is resisted and becomes heat action in the filament, and light is the attendant of intense heat. But, to go back to the sources of electricity.

Frictional electric machines have been constructed in great variety. All, however, embrace the essentials set forth in the sealing-wax experiment, and would be difficult to describe without cuts. Let us, therefore, consider another source of electricity, which was the outgrowth of the discovery of Galvani (or rather his wife), and reduced to concrete form by Volta. We refer to the galvanic or voltaic battery. If we put a bar of zinc into a glass vessel and pour sulphuric acid and water into it, there will be a boiling, and an evolution of hydrogen gas, and energy is released in the form of heat, so that the fluid and the glass vessel become heated. Now let us put a bar of copper or a stick of carbon into the glass, but not in contact with the zinc; connect the ends (that are not immersed) of the two elements--copper and zinc--with a metal wire or any conductor, and a new condition is set up. Heat is no longer evolved to the same extent, but most of the energy becomes electrical in character, and an electrical chain of action takes place in the circuit that has now been formed. Taking the zinc as the starting point, the so-called current flows from the zinc through the fluid to the copper and from the copper through the wire to the zinc.

A chain of polarized atomic activity is established in the circuit, similar to the closed circuit of magnetic lines of force, only the latter is static, while the former is dynamic.

You ask what is the difference? Well, it is much easier to ask a question than it is to answer it. You will remember that in the chapter on magnetism it was stated that the molecules of a magnet were little natural magnets, and that their attractions were satisfied within themselves; that when their local attachments were broken up and all their like poles turned in one direction they could act upon other pieces of iron outside of the magnet. Outside and between the poles there are magnetic lines of force reaching out from one pole to the other. If we put a piece of iron across the poles these lines of force are gathered up and pass through the iron. This is purely a static condition. Let us go back to the cell of battery. When the elements are in position (the copper, the acidulated water and the zinc), and the two wires attached to the two metals which are the two poles of the battery not yet connected, there is a condition induced in these two wires that did not exist before the acidulated water was poured in, although the circuit is not yet established. If we test the two wires we find a difference of potential--a state of strain, so to speak--that did not exist before the acid acted on the zinc and liberated what was stored energy. It is in a static condition, like the magnet, and electrical lines of force are reaching out from both wires so that the ether is in a state of strain between the two poles. The air molecules may partake of it, but we have to bring in the ether as a substance, because the same conditions would practically exist if the two wires were in a vacuum. If now we connect the two wires, we have established a metallic circuit between the two poles of the battery, the static conditions are relieved, the lines of force are gathered up into the wire, and the phenomenon that we call a current is established and we have dynamic or moving electricity.

Having established the so-called electric current we will now try to show you that there really is no current. The idea of a current involves the idea of a fluid substance flowing from one point to another. When you were a boy did you never set up a row of bricks on their ends, just far enough apart so that if you pushed one over they all fell one after another? Now, imagine rows of molecules or atoms, and in your imagination they may be arranged like the bricks, so that they are affected one by the other successively with a rapidity that is akin to that of light-waves, and you can conceive how a motion may be communicated from end to end of a wire hundreds of miles in length in a small fraction of a second, and no material substance has been carried through the wire--only energy. We do not mean to say that the row of bricks illustrates the exact mode of molecular or atomic motion that takes place in a conductor. What we mean is, that in some way motion is passed along from atom to atom.

To give you a better conception of an electric current, let us go back of the galvanic cell to the electric machine. If both poles of the machine are attached to rods terminating in round knobs we can set the machine in action and keep up a steady stream of disruptive discharges that will, if their frequency is great enough, perform the function of a current, and we have dynamic electricity from a statical machine; when the acid of the galvanic battery breaks down a molecule of zinc, energy is set free, and in the battery we have what corresponds to a disruptive discharge of infinitesimal proportions. This discharge would have been immediately converted into heat energy if the copper element had been left out of the battery, but as it is, it impresses itself on the atomic "brick" next to it, which establishes a chain of atomic movement throughout the circuit. This may constitute, if you please, a line of electrical force. But as thousands of these disruptive discharges are taking place simultaneously as many different lines of force are established. You must not conceive of these chains of atoms as simply thrown down like the bricks and left lying there, but that the atom is active; that it has the power to pick itself up again in an infinitesimally short time and is again knocked down (following the illustration of the bricks) by the next discharge along its line or chain of atoms.

If you could get a mental picture of this action you would see that the whole conductor is in a most violent state of atomic motion of a peculiar kind. At the same time a part of this electrical motion is being converted into a heat motion of the atoms, and finally it all returns to heat unless some of it is stored up somewhere as potential energy. If the current has driven a motor that has wound up a weight, a part is stored up in the weight, which has the ability to do work if it is allowed to run down. If it drives machinery as it runs down, the mechanical motion is the expression of the stored energy. When the weight has run down the energy will be represented by the heat created by friction of the journals of the wheels and pulleys and the heating of the air. If the weight is allowed to fall suddenly it will heat the air to some extent, but mostly the earth and the weight itself will be heated. If the source of energy (the battery) is great and the pressure high and the conductor is too small to carry the energy developed in the battery as electricity, heat is developed, and if the heat is sufficiently intense, light also.

We have seen (Vol. II) that heat motion when it reaches a sufficiently high rate throws the ether into a vibratory motion that we call light. However, this vibratory motion of the ether is set up long before it reaches the luminous stage; in other words, there are dark rays of the ether. We find that the electro-atomic motions of a conductor have the power to impress themselves upon the ether.

Let us try another experiment to show that this is the case, not only, but that the impressed ether can transfer these impressions to still another conductor. Suppose we stretch two parallel wires for, say, half a mile, or any distance, only a few feet apart, and make of each a complete circuit by rounding the end of the course and returning the wire to the starting point (as shown in Fig. 1). Put in one of these circuits a battery, and a circuit-breaker (a common telegraph-key), and in the other circuit a galvanometer (an instrument for detecting the presence and measuring the intensity of a galvanic current, by means of a dial and a deflecting needle or pointer). Now if we touch the key and close the circuit in A, the needle of the galvanometer in B will swing in one direction from zero on the dial; and if we release the key, breaking the circuit in A, the needle will swing back in the opposite direction. In neither case will the needle stay deflected, but will at once return to zero.

This shows that when the battery current was allowed to complete its circuit through wire A by closing its key, an electrical action was instantly felt in wire B, although there was no material connection between them other than the air, which is a non-conductor.

The current in the second circuit is called an induced current. Why this current? According to one theory, when we close the primary circuit the surrounding ether is thrown into a peculiar state of strain that we will call magnetic or electrical lines of force. When the ether wave strikes the second wire there is a molecular movement from a state of rest to a state of static strain. During the time that the molecules are moving from the normal to the strained position in sympathy with the ether we have the condition of a dynamic current, which lasts only a moment. This state of strain continues till the circuit is opened (breaking the wire-line), when all the electrical lines of force vanish and the molecular strain of the second wire is relieved, and we again have the conditions, momentarily, for a current of the opposite polarity, and the needle will swing in the opposite direction because the molecules or atoms have, in their recoil to the natural state, moved in an opposite direction.

Going back to Fig. 1, let us further study the phenomena under other conditions. In our first circuit (A) there is a battery and a circuit-breaker, which is a common telegraph-key. Now close the key so that a current will be established. (Remember that "current" is only a name for a condition of dynamic charge.) Place a piece of soft iron across the wire at right angles with the direction of the wire, when of course it will be at right angles with the direction of the current, and you will find now that the iron is more or less magnetic, depending upon the amount of current passing through the wire. If we wind a number of turns of insulated wire through which the current is passing around the iron the magnetism will be increased. In practice there are a certain number of turns and a certain sized wire that will give the best results with a given number of cells of battery (or a given voltage or pressure), operating in a closed circuit of a given resistance. All these questions are worked out mathematically in many standard books on the subject. It is not the intention in these talks to develop the science mathematically but to set out the fundamental physical facts and applications of electricity.

Under the conditions above named magnetism is developed in the soft iron bar. If we open the key the current will cease and the magnetism will vanish-- that is to say, the molecules will turn back to their neutral position by their own attractions, as has been described in a previous chapter. Magnetism developed in this way is called electromagnetism. (See Chap. IV.) If we use a piece of hardened steel instead of the soft iron it will become magnetic and remain so when the circuit is opened, because the natural tendency of the molecules to turn back to the neutral position is not great enough to overcome the coercive force, or molecular friction, of hardened steel, as has been also described in a previous chapter. To make the best electromagnet we need qualities of iron just the opposite from those of the permanent magnet. For the former we need the purest of soft iron, well annealed (heated to redness and slowly cooled, making it less brittle), so that its molecules are free to turn; while for the latter we need hardened steel, so that when the molecules are once wrenched into the magnetic condition they cannot, of themselves, turn back to the neutral state. The great value of the electromagnet lies in its ability to readily discharge, or go back to the neutral state, when the current is broken.

Let us now go back to the beginning of our experiment. When we closed the key and established the current through the wire we found that a piece of iron held at right angles to the wire, although not touching it, became magnetic. We have already said that when the circuit was open, the battery being in circuit, there were electrical lines of force established in the ether, between the two poles of the battery, and that they were gathered up into the conducting wire when the circuit was closed. We now find that there are other lines of force of a different nature established in the ether when the circuit is closed. These we call magnetic lines of force, or the magnetic field of the charged wire, and they are established at right angles to the direction of the current. These magnetic lines of force acting through the ether from an electrically charged conductor are able to break up the natural molecular magnetic rings, referred to in Chapter IV, and turn all their like poles in the same direction--thus making one compound magnet of the iron which in the neutral state consisted of millions of little natural magnets whose attractions were satisfied by a joining of their unlike poles.

Most writers account for all of the phenomena of induced currents in a second wire as coming directly from these magnetic lines of force developed upon closing the circuit.

So much for theory based upon a set of facts that make the theory seem probable. If you don't like it give us a better one. If it is correct the writer claims no credit; it is merely a compilation of suggestions from many sources, including his own experience. We are simply seeking after truth. The man who is an earnest seeker after scientific truth cannot afford to pursue his investigations with any prejudice in favor of one theory more than another, unless the facts sustain him, and then he is not acting from prejudice, but is led by the facts. Many people make pets of their theories; and they become attached to them as they do their children; and they look upon a man who destroys them by a presentation of the facts as an enemy. I once knew a lady who became so attached to her family doctor that, she said, she would rather die under his treatment, if necessary, than to be cured by any other doctor. There are many people who are imbued with this kind of spirit not only in matters scientific, but in matters religious as well. Such people are not the kind who contribute to the world's progress, but are the hindrances that have to be overcome.

CHAPTER VII.

ELECTRIC GENERATORS.

Of the sources of electricity we have mentioned two: Friction, and Galvanism or chemical action. There are hundreds of forms of the latter species of apparatus for generating electrical energy, so we will mention only a few of the more prominent ones. It is not our intention to go into the chemistry of batteries. There are too many exhaustive works on this subject lying on the shelves of libraries that are accessible to all. All galvanic batteries act on one general principle--the generation of electricity by the chemical action of acid on metal plates; but the chemistry of their action is very different. In all batteries the potential energy of one element is greater than the other. The acid of the battery dissolves the element of greater potentiality, and its energy is freed and under right conditions takes on the form of electricity. The potential of zinc, for instance, is greater than that of copper, and the measure of the difference is called the "electromotive force," the unit of which is the "volt." Electromotive force is another name for pressure; the symbol for which is E.M.F.

If we were to put two zinc plates in the battery fluid and connect them in the ordinary way there would be no electricity evolved (assuming that they were perfectly homogeneous), because they are both of the same potential, or have the same possible amount of stored electrical energy measured by its working power. If one of the zinc plates were softer than the other, a feeble current would be developed, for one would be more readily acted upon by the acids than the other. The battery that has been most used in America for telegraphic purposes is called the gravity-battery. It is constructed by putting a copper plate in some form at the bottom of a jar, usually of glass, and filling it partly full of the crystals of sulphate of copper, commonly called "bluestone." Zinc, usually cast in some open form, so as to expose a large surface to the solution, is suspended in the upper part of the jar, which is then filled with water till it covers the zinc. The zinc is the positive metal, but it is called the negative pole. The energy developed by the zinc passes from zinc to copper and out on the circuit from the copper pole. Hence the copper came to be called the positive pole, although in relation to zinc it is negative. Copper would, however, be positive to some other metal whose potential was less. So you see that metals are relative, not absolute, in their character as positive and negative elements.

The galvanic battery has been almost entirely superseded in this country for telegraphic purposes by the dynamo, a machine developing electrical currents by mechanical power. Another form of battery that is extensively used for some kinds of heavy current work is called the storage-battery. The man who did the most, perhaps, to bring the storage-battery to its present state of perfection was Plant? a Frenchman, who died only a short time ago. Although very many types of battery have been developed, it is found that, after all, the lines on which he developed it make the most efficient battery. There is a common notion that electricity is stored in the storage-battery. Energy is stored, that will produce electricity when it is set free, just the same as energy is stored in zinc. The storage-battery, when ready for action, is one form of acid or primary battery. It has been made by passing a current of electricity through it until the chemical relations of the two lead plates have been changed so that the potential of one is greater than that of the other. A simple storage-battery element is made up of two plates of lead held out of contact with each other by some insulating substance the same as the elements of an ordinary battery. The cell is filled with dilute sulphuric acid, and there will be no electrical action till the cell has been charged by running a current of electricity through it and forming a lead oxide on one plate. Now, take off the charging battery and connect the two poles, and electricity will flow until the oxide has partly changed back into spongy metallic lead, when it must be renewed by recharging.

I remember perfectly well the first galvanic battery I ever saw, for it was of my own construction. It is now nearly fifty years ago, and yet it seems but yesterday--such is the flight of time. I related to you in another chapter how I made a voltaic battery--or pile, as it was called--by cutting up my mother's boiler and her stove-zinc, and the domestic incident that followed. Well, a little later I made a real galvanic battery as follows: I lived in the country and far from town or city, and my facilities were extremely limited, so that I pursued my scientific investigations under great difficulties. My only text-book was an old Comstock's Philosophy. In the book was a crude cut of a Morse register and a short description of its construction, including the battery. I determined to make a register, and I did. It was all constructed of wood except the magnet and its armature and the embossing-point, which latter was made of the end of a nail. The thing that seemed out of reach was the electromagnet. I had no money; and there was no one that believed I could do it, and if I could "what good would come of it?" I made friends with a blacksmith by keeping flies off a horse while he nailed the shoes on, and "blowing the bellows" and occasionally using the "sledge" for him. When I thought the obligation had accumulated a sufficient "voltage" (to express it electrically) I communicated to the blacksmith the situation and what I wanted.

The good-natured old fellow was not long in bending up a U magnet of soft iron and forging out an armature. The next step was to wind the U with insulated wire. The only thing that I had ever seen of the kind was an iron wire called "bonnet" wire that was wrapped with cotton thread. This, however, was not available, so I captured a piece of brass bell-wire and wound strips of cotton cloth around it for insulation--and in that way completed the magnet.

Now everything was ready but the battery. I went at its construction with a feeling almost akin to awe, for I could not believe that it would do as described in the book. I procured a candy-jar from the grocer and found some pieces of sheet zinc and copper. These I rolled together into loose spirals and placed one inside the other so that they would not touch, when I was ready for the solution. The druggist trusted me for a half pound of "blue vitriol," and I put it into my battery and filled it with water. I waited awhile for it to dissolve, and then connected my magnet in circuit, when--to my astonishment and delight--it would lift a pound or more. It was a great triumph. I never have had one since that gave me the same satisfaction. But I had my triumph all to myself. I was still the same "tinker" (a name I had long carried), and a nuisance to be endured but not encouraged.

The dynamo is the form of generator now in general use where heavy currents of electricity are needed. It is aptly described by a writer in Modern Machinery, Mr. John A. Grier, as a thing that when "at rest is a lifeless piece of mechanism; in action it has a living spirit as full of mystery as the soul of man." This is a poetic way of describing it that conveys to the mind a sense of the power and beauty of natural law in action, that would not come from a mere recital of the cold scientific facts. The facts, however, are necessary: but let us draw from them all the poetry and all the practical lessons that we can as we go along; for it is this blending of the poetic with the practical that lends a charm to our every-day "grind," and lightens the load of many a weary hour.

The dynamo is a machine that converts mechanical into electrical energy, and the great practical value of energy in this form is that it can be distributed through a conductor economically for many miles. We can transmit mechanical power by means of a rope or cable for a limited distance, but at tremendous loss through friction. We can transmit power through pipes by compressed air or steam, but there is a great loss, especially in the case of steam, by condensation from cold. None of these methods are available for long distances. Another advantage electricity has over other forms of energy is the speed with which it can be transmitted from one place to another. In this respect it has no rival except light. But we have not been able to harness light and make it available to carry either freight or news, except in the latter case for a short distance by flashing it in agreed signals.

The heliostat can be used when the sun shines to transmit news by flashes of sunlight chopped up into the Morse code and thrown from point to point by a moving mirror. But this is limited as to distance; besides, the sun does not always shine. It has the disadvantage in that respect that the old semaphore-telegraph did that was in use in Wellington's day. These semaphores were constructed in various ways, but a common form was that of moving arms that could be seen from hill to hill or point to point. By a code of moving signals news was repeated from point to point and it can be easily imagined that many mistakes occurred, to say nothing of the time it required for repetition. When the battle of Waterloo was fought--so the story goes--news was sent to England by means of the semaphore-telegraph. The dispatch read, "Wellington defeated--" At that point in the message a thick fog came up and lasted for three days, so that no further news could be sent or received. In the telegraphic parlance of to-day the line was "busted." For three long days all London was in deep mourning, when finally the fog lifted, which repaired the telegraphic line, and the balance of the dispatch was received--"the French at Waterloo." Mourning changed to rejoicing and the English have rejoiced ever since when they think of either Wellington or Waterloo.

But to return to the dynamo. The name dynamo is an abbreviation for dynamo-electric machine. A machine for producing dynamic electricity. There are many forms of the dynamo, just as there are in the evolution of every important machine, and there will be many more. But the fundamental, underlying principle of them all is contained in an experiment made by Faraday. Faraday took the soft iron "keeper" of a permanent magnet and wound insulated wire around it and brought the two ends of the wire close together. He now placed the keeper, with the wire wound around it, across the poles of the permanent magnet, and wrenched it away suddenly, when he observed a spark pass between the ends of the wires. This would occur when he approached the poles as well as when he took it away. He discovered that the currents were momentary and occurred at the moment of approach or recession, and that the currents developed by the approach were of opposite polarity to those occurring at the recession. When the "keeper" was put on the poles of the magnet it was magnetized by having its molecular rings broken up and the poles of the little natural magnets all turned in one direction. During the time that the molecules of the keeper are changing they are in a dynamic or moving condition. By some mysterious action of the ether between the iron and the wire wrapped around it there is a corresponding molecular action in the wire that is dynamic for a moment only, and during that moment we have the phenomenon of an electric current. When the magnet and soft iron are separated this molecular state of strain is relieved and the molecules of both the iron and the wire wound about it return to normal, and in the act of returning we have a dynamic or moving condition, resulting in a current, only in the opposite direction. (See Chap. VI.)

Now mount the permanent magnet in a frame and mount the soft iron with the wire on it (which in this shape is an electromagnet) on a revolving arm and so set it on the arm that its ends will come close to, but not touch, the poles of the permanent magnet. Now revolve the arm, and every time the electromagnet or keeper approaches the permanent magnet a current of one polarity will be momentarily developed in the wire of the electromagnet, which is moving. When it is opposite the poles, it has reached the maximum charge and, now, as it passes on it discharges and a current of the opposite polarity is developed in the wire. The more rapidly we revolve the arm the more voltage (electrical pressure) the current it develops will have.

It will be plain to all that we might make the electromagnet stationary and revolve the permanent magnet and get the same result. If the permanent magnet were strong enough and the electromagnet the right size as to iron, windings, etc., and we revolve the arm with sufficient rapidity, we could get an alternating current of electricity that would produce an electric light. I have not and cannot here give you the construction of a modern alternating-current dynamo. I have simply described the simplest form of dynamo, and all of them operate upon the fundamental principle of a permanent magnetic field and an electromagnet, moving in a certain relation to each other. The field may revolve or the electromagnet may revolve, whichever is the most convenient to construct. The field-magnet may be a permanent magnet or an electromagnet, made permanent during the operation of the dynamo by a part of the current generated by the machine being directed through a coil surrounding soft iron; or the field-current may come from an outside source.

This is the kind of field-magnet universally used for dynamo work, as a much stronger magnetism is developed in this way than it is possible to obtain from any system of permanent steel magnets.

The usual construction is to have a stationary field-magnet and then a series of electromagnets mounted and revolving upon a shaft in the center of the magnetic field. The rotating part is called the armature, and is so wound with insulated wire that successive induced currents are created in the armature windings and discharged through brushes which rest on revolving segments that connect with the armature windings. These induced currents succeed each other with such rapidity as to amount in practice to a steady current. However, the separate pulsations are easily heard in any telephone when the circuit is near to that of a dynamo circuit. The dynamo current is not nearly so steady as the battery current, although both are probably made up of separate discharges. In the dynamo there is a discharge every time the electromagnet of the armature cuts through the lines of force of the magnetic field, and in the galvanic battery every time a molecule is broken up and its little measure of energy is set free. In the dynamo the pulsations are so far apart as to make a musical tone of not very high pitch, but in the galvanic battery the pitch of the tone, if there is one, would require a special ear to hear it--one tuned, it may be, up near the rate of light vibration.

There are two types of dynamo, one generating a direct and the other an alternating current. (By alternating we mean first a positive and then a negative current impulse.) We cannot enter into a technical description of the dynamo in a popular treatise such as this.

The dynamo has evolved from the germ discovered by Faraday, till to-day it is a machine, the construction of which requires the highest class of engineering skill. When in action it seems like a great living presence, scattering its energy in every direction in a way that is at once a marvel and a blessing to mankind. But we must not give all the credit to the dynamo. As the moon shines with a reflected light, so the dynamo gives off energy by a power delegated to it by the steam-engine that rotates it, and the steam-engine owes its life to the burning coal, and the burning coal is only giving up an energy that was stored ages ago by the magic of the sunbeam; and the sun--? Well, we are getting close on to the borders of theology, and being only scientists we had better stop with the sun.

There is still another way of generating electricity besides those that we have named; which are friction, chemical action, and the magneto-electric mode of generating a current. Electricity may be generated by heat. If we connect antimony and bismuth bars together and apply heat at the junction of the metals and then connect the free ends of the two bars to a galvanometer, it will indicate a current. These pairs can be multiplied, and in this way increase the voltage or pressure, and, of course, increase the current, if we assume that there is resistance in the circuit to be overcome. If there were absolutely no resistance in the circuit--a condition we never find--there would be no advantage in adding on elements in series.

Substances differ in their resistance to the passage of electricity--the less the resistance the better the conductor. The German electrician, G. S. Ohm (1789-1854), investigated this and propounded a law upon which the unit for resistances is based, and this unit takes his name and is called the "ohm."

Any two metals having a difference of potential will give the phenomena of thermo-electricity. Antimony and bismuth having a great difference of potential are commonly used. The use made of thermal currents is chiefly for determining slight differences of temperature. An apparatus called the thermo-electric pile has been constructed out of a great number of pairs of antimony and bismuth bars. This instrument in connection with a galvanometer makes a most delicate means of determining slight changes of temperature. If one face of a thermopile is exposed to a temperature greater than its own, the needle will move in one direction; if to a temperature lower than its own, the needle will be deflected in the opposite direction. If both faces of the pile are exposed to the same changes of temperature simultaneously, of course no electrical manifestations will occur.

The earth is undoubtedly a great thermal battery that is kept in action by the constant changes of temperature going on at the earth's surface, caused by its rotation every twenty-four hours on its axis. The sun, of course, is at some point heating the earth, which at other points is cooling, making a constant change of potential between different points. If we heat a metal ring at one point a current of electricity will flow around it--especially if it is made of two dissimilar metals--until the heat is equally distributed throughout the ring.

Some years ago, when the Postal Telegraph Company first began operations between New York and Chicago, the writer made observations twice a day for some time of the temperature and direction of the earth-current. The first two wires constructed gave only two ohms resistance to the mile, which facilitated the experiments. I found that in almost every instance the current flowed from the point of higher temperature to the lower. If the temperature in New York were higher at the time of observations than in Chicago the current would flow westward, and if the conditions were reversed the current would be reversed also.

CHAPTER VIII.

ATMOSPHERIC ELECTRICITY.

Nature has another mode of generating electricity, called atmospheric. The normal conditions of potential between the earth and the upper atmosphere seem to be that the atmosphere is positively electrified and the earth negatively. These conditions change, apparently from local causes, for short periods during storms. In some way the sun's rays have the power directly or indirectly to give the globules of moisture in the air a potential different from that of the earth.

In clear weather we find the air near to the earth in a neutral condition, but gradually assuming the condition of a positive charge as we ascend; so that the upper air and the earth are oppositely charged like the two sides of a Leyden jar or two leaves of a condenser. This condition is intensified and localized when a thunder-cloud passes over the earth. The moisture globules have been charged with potential energy by the power of the sun's rays when evaporation took place; but in this state the energy is neither heat nor electricity, but a state of strain like a bent bow or a wound-up spring. When these moisture globules condense into drops of water the potential energy is set free and becomes active either as heat or electricity. The cloud gathers up the energy into a condensed form, and when the tension gets too great a discharge takes place between the cloud and the earth or from one cloud to another, which to an extent equalizes the energy.

In most cases of thunder and lightning it is only a discharge from cloud to cloud unequally charged. This does not relieve the tension between the earth and the cloud, but distributes it over a larger area. The reason for this constant electrical difference between the earth and the upper regions of atmosphere is not well understood, except that primarily it is an effect of the sun's rays. Evaporation may and probably does play a part, and the same causes that give rise to the auroral display may contribute in some way to the same result. Evaporation does not always take place at the earth's surface. Cloud formations may be evaporated in the upper air into invisible moisture spherules, and charged at the time with potential energy. If we go up into a high mountain when the conditions are right, we can witness the effect of this condition of electrical charge or strain between the upper regions of the atmosphere and the earth, and the tendency to equalize the potentials between the clouds and the earth. Often one's hair will stand on end, not from fright, but from electricity passing down from the upper regions to the earth. When the tension is very great a loud hissing sound as of many musical tones of not very good quality may be heard, and a brush or fine-pointed radiation of electricity may be seen from every point, even from your finger-ends. The thunder is not usually so loud on high mountains for two reasons-- one because the air is rare, but the chief reason is that the mountain acts as a great lightning-rod and gradually discharges the cloud or atmosphere, for often the phenomena may occur when the sky is clear.

I remember being on top of what is called the Mosquito Range, between Alma and Leadville in Colorado, during the passage of a thunder shower. There was no heavy thunder, but a constant fusillade of snapping sounds, accompanied by flashes not very intense. I could feel the shocks, but not painfully. A part of the time I was in the cloud and became for the time being a veritable lightning-rod. After the cloud passed it crawled down the mountainside as if clinging to it, all the time bombarding it with little electric missiles. After the cloud left the mountain and passed over the valley I could hear loud thunder, because the charge would have to accumulate quite a quantity, so to speak, before it could discharge. These heavy discharges when the cloud is some distance from the earth would be dangerous to life, while the light ones, when the cloud is in contact with the earth, are not.

Many wonderful and destructive effects come from these lightning discharges and many lives are lost every year from this cause. I do not suppose it is possible to be on one's guard continually, but many lives are needlessly lost either from ignorance or carelessness. Although there is a just prejudice against lightning-rods as ordinarily constructed, it is still just as possible to protect your house and its inmates from the destroying effects of lightning as from rain. If, for instance, we lived in metal houses that had perfect contact all round them with moist earth, or better, with a water-pipe that has a large surface contact with the earth, the lightning would never hurt the house or the inmates. In such a case you simply carry the surface of the earth to the top of your house, electrically speaking, and neutralization takes place there in case the lightning strikes the house. A house that is heated with hot water can easily be made lightning-proof by a little work at the top and bottom of the heating system. All the heavy metal of the house should be a part of the lightning-rod. Points should be erected at the chimneys, and if there is a metal roof they should be connected with it. Then connect the roof with rods from several points with the ground. Here is where most rods fail. The ground connection is not sufficient. The earth is a poor conductor, and we have to make up by having a large metal surface in contact with it. It is best to have the rod connected with the water pipe, if there is one, and have it connected with metal running all around the house as low down as the bottom of the cellar, for sometimes there is an upward stroke, and you never can tell where it is coming up. If you have a heating system it should be thoroughly grounded and the top pipe connected with the rods at the chimneys. These rods need not be insulated as is the usual practice.

If you are outdoors during a thunder-storm never get under a tree, but if you are twenty or thirty feet away it may save your life, because, if it comes near enough to strike you, it will probably take the tree in preference. It seeks the earth by the easiest passage. An oil-tank and a barn are dangerous places, if the one has oil in it and the other is filled with hay and grain. A column of gas is rising that acts as a conductor for lightning. Of course if the barn is properly protected with rods it will be safe. Sometimes a cloud is so heavily charged that the lightning comes down like an avalanche, and in such a case the rods must have great capacity and be close together to fully protect a building.

There is a popular notion that rods draw the lightning and increase the damage rather than otherwise. This is a mistake. Points will draw off electricity from a charged body silently. It would be possible to so protect a district of any size in such a way that thunder would never be heard within its boundaries if we should erect rods enough and run them high enough into the upper air. The points--if they were close enough together--would silently draw off the electricity from a cloud as fast as it formed, and thus effectually prevent any disruptive discharge from taking place.

CHAPTER IX.

ELECTRICAL MEASUREMENT.

Having given a short account of some of the sources of electricity, let us now proceed to describe some of the practical uses to which it is put, and at the same time describe the operation of the appliances used. Before proceeding further, however, we ought to tell how electricity is measured. We have pounds for weight, feet and inches for lineal measure, and pints, quarts, gallons, pecks and bushels for liquid and dry measure, and we also have ohms, volts, amperes and ampere-hours for electricity.

When a current of electricity flows through a conductor the conductor resists its flow more or less according to the quality and size of the conductor. Silver and copper are good conductors. Silver is better than copper. Calling silver 100, copper will be only 73. If we have a mile of silver wire and a mile of iron wire and want the iron wire to carry as much electricity as the silver and have the same battery for both, we will have to make the iron wire over seven times as large. That is, the area of a cross-section of the iron wire must be over seven times that of the silver wire. But if we want to keep both wires the same size and still force the same amount of current through each we must increase the pressure of the battery connected with the iron wire. We measure this pressure by a unit called the "volt," named for Volta, the inventor or discoverer of the voltaic battery. The volt is the unit of pressure or electromotive force. (In all these cases a "unit" is a certain amount or quantity--as of resistance, electromotive force, etc.--fixed upon as a standard for measuring other amounts of the same kind.)

The iron wire offers a resistance that is about seven times greater than silver to the passage of the current. To illustrate by water pressure: If we should have two columns of water, and a hole at the bottom of each column, one of them seven times larger than the other, the water would run out much faster from the larger hole if the columns were the same height. Now, if we keep the column with the larger hole at a fixed height a certain amount of water will flow through per second. If we raise the height of the column having the small hole we shall reach a point after a time when there will be as much water flow through the small hole per second as there is flowing through the large hole. This result has been accomplished by increasing the pressure. So, we can accomplish a similar result in passing electricity through an iron wire at the same rate it flows through a silver wire of the same size, by increasing the pressure, or electromotive power; and this is called increasing the voltage.

The quality of the iron wire that prevents the same amount of current from flowing through it as the silver is called its resistance. The unit of resistance, as mentioned in the last chapter, is called the ohm, and the more ohms there are in a wire as compared with another, the more volts we have to put into the battery to get the same current.

The unit for measuring the current is called the "ampere," named after the French electrician, A. M. Ampere (1789-1836).

Now, to make practical application of these units. The volt is the potential or pressure of one cell of battery called a standard cell, made in a certain way. The electromotive force of one cell of a Daniell battery is about one volt. One ohm is the resistance offered to the passage of a current having one volt pressure by a column of mercury one millimeter in cross-section and 106.3 centimeters in length. Ordinary iron telegraph-wire measures about thirteen ohms to the mile. Now connect our standard cell--one volt--through one ohm resistance and we have a current of one ampere. Unit electromotive force (volt) through unit resistance (ohm) gives unit of current (ampere). It is not the intention to treat the subject mathematically, but I will give you a simple formula for finding the amount of current if you know the resistance and the voltage. The electromotive force divided by the resistance gives the current. $C = E/R$ or current (amperes) equals electromotive force (volts) divided by the resistance (ohms).

But still further: One ampere of current having one volt pressure will develop one watt of power, which is equal to 1/746 of a horse-power. (The watt is named in honor of James Watt, the Scottish inventor of the steam-engine--1786-1813). In other words, 746 watts equal one horse-power. By multiplying volts and amperes together we get watts.

If we want to carry only a small current for a long distance we do not need to use large cells, but many of them. We increase the pressure or voltage by increasing the number of cells set up in series. If we have a wire of given length and resistance and find we need 100 volts to force the right amount or strength of current through it, and the electromotive force of the cells we are using is one volt each, it will require 100 cells. If we have a battery that has an E. M. F. of two volts to the cell, as the storage-battery has, fifty cells would answer. If we want a very strong current of great volume, so to speak, for electric light or power, and use a galvanic battery, we should have to have cells of large surface and lower resistance both inside and outside the cells.

When dynamos are used they are so constructed that a given number of revolutions per minute will give the right voltage. In fact, the dynamo has to be built for the amount of current that must be delivered through a given resistance. The same holds good for a dynamo as for a galvanic battery. If any one factor is fixed, we must adapt the others to that one in order to get the result we want. There are many other units, but to introduce them here would only confuse the reader. The advanced student is referred to the text-books.

With this much as a preliminary we are prepared to take up the applications of electricity, which to most people will be more interesting than what has gone before.

CHAPTER X.

THE ELECTRIC TELEGRAPH.

In the year 1617 Strada, an Italian Jesuit, proposed to telegraph news without wires by means of two sympathetic needles made of loadstone so balanced that when one was turned the other would turn with it. Each needle was to have a dial with the letters on it. This would have been very nice if it had only worked, but it was not based on any known law of nature.

Many attempts at telegraphing with electricity were made by different people during the eighteenth century. About 1748 Franklin succeeded in firing spirits by means of a wire across the Schuylkill River, using, as all the other experimenters had done, frictional electricity. In 1753 an anonymous letter was written to Scott's Magazine describing a method by which it was possible to communicate at a distance by electricity. The writer proposed the use of a wire for each letter of the alphabet, that should terminate in pith balls at the receiving end, and under the balls were to be strips of paper corresponding to the letters of the alphabet. The message was to be sent by discharging static electricity through the wire corresponding to the first letter of a word when the paper would be attracted to the pith ball and read by the observer. Then the wire corresponding to the second letter of the word was to be charged in like manner, and so on till the whole message was spelled out. This was the first practical (i.e., possible) suggestion for a telegraph. The writer also proposed to have the wires strung on insulators, which was a great advance over the other attempts.

The communication was anonymous, as no doubt, like many others, the author feared the ridicule of his neighbors. It requires a vast amount of moral courage to stand up before the world and openly advocate some new theory that has never come within the experience of any one before. It requires much now, but it required more then; for a man in those days would have been roasted for what in these days he would be toasted. The rank and file of humanity have been opposed to innovations in all ages, but no progress could have been made without innovations. There always has to be a first time. Galileo is said to have been forced to retract, on his knees, some theory he advanced about the motion of the earth, and its relation to the sun and other heavenly bodies. Notwithstanding this retraction the seed-thought sown by Galileo took root in other minds, which led to the triumph of scientific truth over religious fanaticism.

The writer in Scott's Magazine did not have the opportunity to put his ideas into practice, so the glory of the invention fell to others. Such men as this unknown writer are made of finer stuff, and they stand alone on the frontier of progress. They do not fear the bullets of an enemy half so much as the gibes of a friend. Much of their work is done quietly and without notice, and when something of real importance is worked out theoretically and experimentally, some one seizes upon it and proclaims it from the housetops and attaches to it his name; but perhaps years after the real inventor (the man who taught the so-called inventor how to do it) is dead, some one writes a book that reveals the truth, and then the hero-loving people erect a monument to his memory.

Such a man was our own Professor Joseph Henry, so long the presiding genius at the Smithsonian Institution at Washington. He worked out all the problems of the present American telegraphic system and demonstrated it practically. Everything that made the so-called Morse telegraph a success had long before been described and demonstrated by Henry. Yet with the modest grace that was ingrained in the man he yielded all to the one who was instrumental in constructing the first telegraph line between Baltimore and Washington. Great credit is due to such men as Morse and Cyrus W. Field-- neither of them inventors, but promoters of great systems of communication that are of unspeakable benefit to mankind. Henry pointed out the way, and Morse carried it into effect. Morse has had no more credit than was due him, but has Henry had as much as is due him? No great invention was ever yet the work, wholly, of one man. We Americans are too apt to forget this.

I shall always remember Henry as a most unassuming, kindly, genial man, and I shall never forget his kindness to me. In 1874 I began my researches in telephony, having applied for a patent for an apparatus for transmitting musical tones telegraphically. This consisted of a means of transmitting musical tones through a wire and reproducing them on a metal plate (stretched on the body of a violin to give it resonance) by rubbing the plate with the hand--the latter being a part of the circuit. The examiner refused the application at first on the ground that the inventor or operator could not be a part of his machine. I took my apparatus and went to Washington, first calling upon Professor Henry, never having met him before. He received me most kindly, and allowed me to string wires from room to room in the institute, and when he had witnessed the experiments he seemed as delighted as a child. I now brought the patent office official over to the Smithsonian and soon convinced him that the inventor could be a part of his own machine.

The same year I went abroad, and Henry gave me a letter to Tyndall. It was very fortunate for me that he did, for Tyndall was very shy at first, and it was only Henry's letter that gave me a hearing for a moment. The history of the few days that followed this first interview with Tyndall at the Royal Institution would make very interesting reading, if I felt at liberty to publish it. Suffice it to say that he was convinced in a few minutes after he had reached the experimental stage that not all my work had been anticipated by Wheatstone, as he asserted before seeing the experiments. Wheatstone had transmitted the tones of a piano, mechanically, from one room to another by a wooden rod placed upon the sound-board and terminating in another room in contact with another sound-board. But this was very different from transmitting musical tones and melodies from one city to another through a wire, as I could do with my electrotelephonic apparatus.

It is a curious fact that the world is divided into two great classes, leaders and followers. Or we might say, originators and copyists; the former division being very small, while the latter is very large. As late as 1820 the European philosophers were trying to construct a telegraphic system based upon two ideas, announced a long time before, to wit, the use of static or frictional electricity, and a wire for every letter. It does not seem to have occurred to any one to devise a code consisting of motions differently related as to time, and to use simply one wire.

In 1819 Oersted discovered the effect of a galvanic current on a magnetic needle, and published a pamphlet concerning his discovery. This stimulated others, and Ampere applied it to the galvanometer the same year. Arago applied it to soft iron, and here was the germ of the electromagnet. We see that as far back as 1820 we had the galvanic battery and the electromagnet, the two great essentials of the modern telegraph.

However, there remained another great discovery to be made before these elements could be utilized for telegraphic purposes. One cell of battery was used, and the magnet was made by winding one layer of wire spirally around the iron, so that each spiral was out of touch with its neighbor. Barlow of England, a Fellow of the Royal Society, tried the effect of a current through a wire 200 feet long, and found that the power was so diminished that he was discouraged, and in a paper gave it as his opinion that galvanism was of no use for telegraphing at a distance. This paper stimulated others, and it was reserved for our own Joseph Henry, already referred to, to show not only how to construct a magnet for long-distance telegraphy, but also how to adapt the battery to the distance. He showed us that by insulating the wire and using several layers of whirls, instead of one, and by using enough cells of battery coupled up in series to get more voltage, as we now express it, it was possible to transmit signals to a distance. He not only set forth the theory, but he constructed a line of bell-wire 1060 feet long and worked his magnet by making the armature strike a bell for the signals, which is the basis of the modern "sounder."

Nothing was needed but to construct a line and devise a code to be read by sound, to have practically our modern Morse telegraph. This line was constructed in 1831. In 1835 Henry, who was then at Princeton, constructed a line and worked it as it is to-day worked, with a relay and local circuit, so that at that period all the problems had been worked out. But, like the speaking-telephone in its early inception, no one appreciated its real importance. Henry himself did not think it worth while to take out a patent. Two years later the Secretary of the Treasury sent out a circular letter of inquiry to know if some system of telegraphic communication could not be devised. The learned heads of the Franklin Institute of Philadelphia, the oldest scientific society in America, advised that a semaphore system be established between New York and Washington, consisting of forty towers with swinging arms, the same as used in the days of Wellington. Among other replies to the circular letter of the secretary was one from Samuel F. B. Morse. Morse was not a scientist or even an inventor, at least not at that time. He was an artist of some note. In 1832, while crossing the ocean, Morse, in connection with one Dr. Jackson of Boston, devised a code of telegraphic signs intended to be used in a chemical telegraph system.

Some years later Morse adapted Henry's signal-instrument to a recorder, called the Morse register, and this was the instrument used in the early days of the Morse telegraph.

What Morse seems to have really invented was the register, which made embossed marks on a strip of paper, and the code of dots and dashes representing letters, now known as the Morse alphabet, although this latter is questioned. Morse took his apparatus to Washington and exhibited it to the members of Congress in the year 1838, but it was four years before a bill was passed that enabled him to try the experiment between Baltimore and Washington. We will let him describe in his own words the closing day of Congress. He says:

"My bill had indeed passed the House of Representatives and it was on the calendar of the Senate, but the evening of the last day had commenced with more than 100 bills to be considered and passed upon before mine could be reached. Wearied out with the anxiety of suspense, I consulted one of my senatorial friends. He thought the chance of reaching it to be so small that he advised me to consider it as lost. In a state of mind which I must leave you to imagine, I returned to my lodgings to make preparations for returning home the next day. My funds were reduced to the fraction of a dollar. In the morning, as I was about to sit down to breakfast, the servant announced that a young lady desired to see me in the parlor. It was the daughter of my excellent friend and college classmate, the commissioner of patents, Henry L. Ellsworth. She had called, she said, by her father's permission, and in the exuberance of her own joy, to announce to me the passage of my telegraph bill at midnight, but a moment before the Senate adjourned. This was the turning-point of the telegraph invention in America. As an appropriate acknowledgment of the young lady's sympathy and kindness--a sympathy which only a woman can feel and express--I promised that the first dispatch by the first line of telegraph from Washington to Baltimore should be indited by her; to which she replied: 'Remember, now, I shall hold you to your word.' About a year from that time the line was completed, and, everything being prepared, I apprised my young friend of the fact. A note from her inclosed this dispatch: 'What hath God wrought?' These were the first words that passed on the first completed line in America."

   The first telegraph-line in America was put into operation in the spring of 1844 at the beginning of Polk's administration. I remember as a boy having the two cities, Baltimore and Washington, pointed out to me on the map, and how the story of the telegraph impressed me. Congress appropriated $30,000 for the construction of the line, and $8000 to keep it running the first year. It was placed under the control of the postmaster-general, and the line was thrown open to the public. The tariff was fixed at one cent for every four words. It was open for business on April 1, 1844, and for the first few days the revenue was exceedingly small. On the morning of the first day a gentleman came in and wanted to "see it work." The operator told him that he would be glad to show it at the regular tariff of one cent for four words. The gentleman grew angry and said that he was influential with the administration, and that if he did not show him the working free of charge he would see to it that he lost his job. His bluff did not succeed. The operator referred him to the postmaster-general, and thus the stormy interview ended. No patrons came in for the next three days, but a great number stood around hoping to see the instrument start up, but no one was willing to invest a cent-

-probably from fear of being laughed at.

On the fourth day the same gentleman who had threatened the young man with dismissal came back and invested a cent, and this was the first and only revenue for four days. The message that was sent only came to one-half cent, but as the operator could not make change the stranger laid down the cent and departed. His name ought to be known to fame as the first man patron of the telegraph.

The operation of the Morse telegraph is very simple if we grant all that has gone before. All that is needed is the wire, the battery, and the key, as shown in Fig. 2 (page 99), and a relay--an extra electromagnet which receives the electric current and by its means puts into or out of action a small local battery on a short circuit in which is placed the receiving or recording apparatus. Thus we have a wire starting from the earth in New York and passing through a battery, a key and a relay, and thence to Boston on poles, with insulators on which the wire is strung, and through another instrument, key and battery in Boston, the same as at the New York end, and into the ground, leaving the earth to complete one-half of the circuit. When the keys at both ends are closed the batteries are active and the armatures or "keepers" are attracted so that the armature levers rest on the forward stops. (See diagram Fig. 2.) If either one of the keys is opened the current stops flowing and the magnetism vanishes from all the electromagnets on the line, and a spring or retractile of some kind pulls the armatures away from the magnets and the levers rest on their back stops. In this way all the levers of all the magnets are made to follow the motions of any key. If there are more than two magnets in circuit (and there may be twenty or more) they all respond in unison to the working of one key, so that when any one station is sending a dispatch all the other stations get it.

But there is a "call" for each office, so that the operator only heeds the instrument when he hears his own call. Operators become so expert in reading by sound that they may lie down and sleep in the room, and, although the instrument is rattling away all the time, he does not hear it till his own call is made, when he immediately awakes.

In the old days messages were received on slips of paper by the Morse register by means of dots and dashes. Gradually the operator learned to read by sound, till now this mode of receiving is almost universal the world over. Reading by sound was of American origin. It is a spoken language, and when one becomes accustomed to it it is like any other language. This code language has some advantages over articulate speech, as well as many disadvantages. A gentleman who was connected with a Louisville telegraph office told me that one of the best operators he ever knew was as deaf as a post. He would receive the message by holding his knee against the leg of the table upon which the sounder was mounted, and through the sense of feeling receive the long and short vibrations of the table, and by this means read as well or better than through the ear, because he was not distracted by other sounds.

A story is told of the late General Stager that at one time he was on a train that was wrecked at some distance from any station. He climbed a telegraph pole, cut the wire and by alternately joining and separating the ends sent a message, detailing the story of the wreck, to headquarters, and asked for assistance. He then held the two ends of the wire on each side of his tongue and tasted out the reply--that help was coming. Any one who has ever tasted a current knows that it is very pronounced.

A story similar to this is told of the early days when the Bain chemical system was used between Washington City and some other point. This system made marks on chemically-prepared paper; as the current passed through it left marks on the paper from the decomposition of the chemicals. Some of the preparations emitted an odor during the time that the current passed. The occurrence to which we refer took place at presidential election time. At some station out of Washington an operator was employed who had a blind sister, and this sister knew the Morse alphabet well before she became blind. One evening a signal came to get ready for a message containing the returns from the election. In the hurry, and just as the message had started, the lamp was upset and they were in total darkness--at least, the brother was. The sister, poor girl, had been in darkness a long time. The blind sister leaned over the stylus through which the current flowed to the paper and smelled out as well as spelled out the message, and repeated it to her astonished brother.

By the old semaphore system the motions were sensed through the eye as well as the early method of cable signaling. It will be seen from the above that the Morse code may be communicated through any one of the five senses.

CHAPTER XI.

RECEIVING MESSAGES.

With but few exceptions the Morse code is the one almost universally used the world over. As it is used in Europe, it is slightly changed from our American code, but they all depend upon dots, dashes, and spaces, related in different combinations, for the different letters. Notwithstanding its universal use it is not free from serious difficulties in transmission unless it is repeated back to the sender for correction; and then in some cases it is impossible to be sure, owing to difficulties of punctuation and capitalizing, and the further difficulty of running the signals together, caused, it may be, by faulty transmission, induced currents from other wires, "swinging crosses" or atmospheric electricity. Sometimes it is a psychological difficulty in the mind of the receiving-operator. The telegraph companies have to suffer damages from all these and many other unforeseen causes.

Prescott tells some curious things that happened in the early days, growing out of the peculiarities of the receiving-operator. At one time he was reporting by telegraph one of Webster's speeches made at Albany in 1852 in which there were many pithy interrogative sentences, and he was desirous of having the interrogation-points appear. So to make sure, whenever he wished an interrogation-point he said "question" at the end of almost every sentence. Next day he was horrified on reading the speech to see the ends of the sentences bristling with the word "question."

Some time back in the fifties a gentleman in Boston telegraphed to a house in New York to "forward sample forks by express." The message when received by the New York merchant read: "Forward sample for K. S. by express." The New York merchant did not know who K. S. was, nor did he gather from the dispatch what kind of sample he wanted. So he went to the telegraph office to have the matter cleared up. The Boston operator repeated the message, saying "sample forks." "That's the way I received it and so delivered it--sample for K. S.," said New York. "But," says Boston, "I did not say for K. S.; I said f-o-r-k-s." New York had read it wrong in the start and could not get it any other way. "What a fool that Boston fellow is. He says he did not say for K. S., but for K. S." Boston had to resort to the United States mail before the mystery was solved.

Curiously enough, the old method of recording the dots and dashes on the paper strip was not so reliable as the present mode of reading by sound. A man can put his individuality to some extent into a sounder, and when one becomes used to his style it is much easier to read him accurately by sound than by the paper impressions. Some people never could learn to read either by paper or sound. An instance of this kind is given of a middle-aged man who was employed by a railroad company as depot master and telegraph operator, in the old days of the paper strip. One day he rushed out and hailed the conductor of a train that had just pulled into the station, and told him that ---- train had broken both driving-wheels and was badly smashed up. The conductor could read the mystic symbols, so he took the tape and deciphered the dispatch as follows: "Ask the conductor of the Boston train to examine carefully the connecting-rods of both driving-wheels, and if not in good condition to await orders." It is further related of this same operator that when he got into real difficulty with his "tape" he used to run over to the regular commercial office to have his messages translated. One day he rushed into his neighbor's office trailing the tape behind him and saying: "I am sure an awful accident has happened by the way the message was rattled off." A playful dog had torn off a large part of the strip as it trailed along, so only a part was left. It read, "Good morning, Uncle Ben. When are you----" The dog had swallowed the balance of the dispatch.

Sometimes the Morse code is not only funny but disastrous. A gentleman wanted to borrow money of some capitalists who, not knowing his financial standing, telegraphed to a banker who they knew could post them. They received an answer, "Note good for large amount." The gentleman borrowed a "large amount," but afterward when it came to be investigated it was found that the dispatch was originally written "not," instead of "note," which made "all the difference in the world."

It has been stated that any one of the five senses may be called into service to interpret the Morse code into words and ideas. A story is told by Mr. Prescott that he says is true, as he knew the party. A friend of his, by name Langenzunge, who knew the Morse code, had served under General Taylor (who at this time was President) at Palo Alto, in Mexico. The general had just promised him an office; soon after he left Washington for the west over the Baltimore and Ohio on a freight train; the President was taken seriously ill, and his friend hearing of it was troubled not only because he loved the old general, but on account of the change in his own prospects. The train stopped somewhere on the Potomac at midnight and remained there for four hours. Uneasy and sad, he wandered down the track and climbed a pole, cut the wire and placed the ends each side of his tongue and tasted out the fatal message--"Died at half-past ten." The shock (not the electric) was so great that he almost fell from the pole.

What a situation! A man climbs a pole at midnight miles from the sick friend he loves, puts his tongue to inanimate wire, and is told in concrete language-- through the sense of taste--that his friend is dead. This is only one of the many, many wonderful episodes of the telegraph.

CHAPTER XII.

MISCELLANEOUS METHODS.

"It never rains but it pours." Almost simultaneously with the demonstration of the Morse telegraph other types were devised. There were the needle systems of Cooke and Wheatstone, the chemical telegraph of Alexander Bain, and soon the printing telegraph of House, and later that of Hughes. The latter is in use on the continent of Europe, and a modification of it has a very limited use on some American lines. The Bain telegraph used a key and battery the same as the Morse system, but it did not depend upon electromagnetism as the Morse system does. When in operation a strip of paper was made to move under an iron stylus at the receiving-end of the line. The paper was saturated with some chemical that would discolor by the electrolytic action of the current. When a message was sent the paper was set to moving by a clock mechanism or otherwise, under the stylus that was pressing on the paper as it passed over a metal roller or bed-plate. The transmitting-operator worked his key precisely as in sending an ordinary message by the Morse system. The effect was to send currents through the receiving-stylus chopped into long or short marks, or the dots and dashes of the Morse code, and recorded on the tape in marks that were blue or brown, according to the chemical used. A few lines were established in this country on the Bain system, but it never came into general use.

A number of systems, called "automatic," grew out of the Bain system. Bain himself devised, perhaps, the first automatic telegraph. The fundamental principle of all automatic telegraphs depends upon the preparation of the message before sending, and is usually punched in a strip of paper and then run through between rollers that allow the stylus to ride on the paper and drop through the holes that represent the dots and lines of the Morse alphabet. Every time the stylus drops through a hole in the paper it makes electrical contact and sends a current, long or short, according to the length of the hole. The object of the automatic system was to send a large amount of business through a single wire in a short time. It does not save operators, as the messages have to be prepared for transmission, and then translated at the receiving-end and put into ordinary writing for delivery.

The automatic system is not used except for special purposes, and the one that seems to be the most favored is that of Wheatstone. The system is in use in England and in America to a limited degree.

Early in the history of the telegraph a printing system was devised. Wheatstone and others had proposed systems of printing telegraphs in Europe, but these never passed the experimental stage. The first printing telegraph introduced in America was invented by Royal E. House of Vermont, and first introduced in 1847 on a line between Cincinnati and Jeffersonville, a distance of 150 miles. In 1849 a line for commercial use was established between New York and Philadelphia, and for some years following many lines were equipped with the House printing telegraph instrument. The late General Anson Stager was a House operator at one time. All printing telegraph instruments, while differing greatly in detail, have certain things in common, to wit: a means for bringing the type into position, an inking device, a printing mechanism, a paper feed, and a means for bringing the type-wheels into unison. There are two general types of printing instruments, the step-by-step, and the synchronously moving type-wheels. The House printer was a step-by-step instrument and consisted of two parts, a transmitter and a receiver. The transmitter consists of a keyboard like a piano, with twenty-eight keys. These keys are held in position by springs. Under the keys is a cylinder having twenty-eight pins on it corresponding to the twenty-six letters of the alphabet and a dot and a space. This cylinder was driven by some power. In those days it was by man-power. It was carried by a friction, so that it could be easily stopped by the depression of any one of the keys that interfered with one of the pins. One revolution of the cylinder would break and close the current twenty-eight times, making twenty-eight steps.

The receiving-instrument consisted of a type-wheel and means for driving it. It was somewhat complicated, and can only be described in a general way. If the cylinder of the transmitter was set to rotating it would break and close twenty-eight times each revolution. (There were fourteen closes and fourteen breaks, each break and each close of the current representing a step.) The type-wheel of the receiver was divided into twenty-eight parts, having twenty-six letters and a dot and space, each break moved it one step and each close a step; so that if the cylinder, with its twenty-eight pins, started in unison with the type-wheel, with its twenty-eight letters and spaces, they would revolve in unison. The keys were lettered, and if any one was depressed the pin corresponding to it on the cylinder would strike it and stop the rotation of the cylinder, which stopped the breaking and closing of the circuit, which in turn stopped the rotation of the type-wheel--and not only stopped it, but also put it in a position so that the letter on the type-wheel corresponding to the letter on the key that was depressed was opposite the printing mechanism. The printing was done on a strip of paper, which was carried forward one space each time it printed. The printing mechanism was so arranged that so long as the wheel continued to rotate it was held from printing, but the moment the type-wheel stopped it printed automatically.

The messages were delivered on strips of paper as they came from the machine.

In 1855 David E. Hughes of Kentucky patented a type-printing telegraph that employed a different principle for rotating the type-wheel. The electric current was used for printing the letters and unifying the type-wheels with the transmitting-apparatus. The transmitter, cylinder, and the type-wheel revolved synchronously, or as nearly so as possible, and the printing was done without stopping the type-wheel. Whenever a letter was printed the type-wheel was corrected if there was any lack of unison.

This type of machine in a greatly improved form is still used on some of the

Western Union lines, especially between New York, Boston, Philadelphia, and Washington. It is also in use in one of its forms in most of the European countries.

CHAPTER XIII.

MULTIPLE TRANSMISSION.

Although the printing and automatic systems of telegraphing are used in America to some extent, the larger part is done by the Morse system of sound-reading and copying from it, either by pen or the typewriter. In the early days only one message could be sent over one wire at the same time, but now from four to six or even more messages may be sent over the same wire simultaneously without one message interfering with the other. Like most other inventions, many inventors have contributed to the development of multiple transmission, till finally some one did the last thing needed to make it a success. The first attempts were in the line of double transmission, and many inventors abroad have worked on this problem.

Moses G. Farmer of Salem, Mass., proposed it as early as 1852, and patented it in 1858. Gintl, Preece, Siemens and Halske and others abroad had from time to time proposed different methods of double transmission, but no one of them was a perfect success. When the line was very long there was a difficulty that seemed insurmountable. In the common parlance of telegraphy, there was a "kick" in the instrument that came in and mutilated the signals. About 1872 Joseph B. Stearns of Boston made a certain application of what is called a "condenser" to duplex telegraphy that cured the "kick," and from that time to this it has been a success. Farther along I will tell you what occasioned this "kick" and how it was cured. If this or some other method could be applied as successfully to cure the many chronic "kickers" in the world it would be a great blessing to mankind.

It has always been a mystery to the uninitiated how two messages could go in opposite directions and not run into one another and get wrecked by the way. If you will follow me closely for a few minutes I will try to tell you.

We have already stated that an electromagnet is made by winding an insulated wire around a soft iron core. If we pass a current of electricity through this wire the core becomes magnetic, and remains so as long as the current passes around it. In duplex telegraphy we use what is called a differential magnet. A differential electromagnet is wound with two insulated wires and so connected to the battery that the current divides and passes around the iron core in opposite directions. Now if an equal current is simultaneously passed through each of the wires of the coil in opposite directions the effect on the iron will be nothing, because one current is trying to develop a certain kind of polarity at each pole of the magnet, while the current in the other wire is trying to develop an opposite kind in each pole. There is an equal struggle between the two opposing forces, and the result is no magnetism. This assumes that the two currents are exactly the same strength.

If we break the current in one of the coils we immediately have magnetism in the iron; or if we destroy the balance of the two currents by making one stronger than the other we shall have magnetism of a strength that measures the difference between the two.

Without specifically describing here the entire mechanism--since this is not a text-book or a treatise--we may say that a duplex telegraph-line is fitted with these differentially wound electromagnets at every station. When Station A (Fig. 3) is connected to the line by the positive pole of its battery, Station B will have its negative pole to line and its positive to earth. When A depresses his key to send a message, half the current passes by one set of coils around his differential magnet through a short resistance-coil to the earth, and the other half by the contrary coil around the magnet to the line, and so to Station B. The divided current does not affect A's own station, being neutralized by the differential magnet, but it does affect B, whose instrument responds and gives him the message.

Now B may at the same time send a message to A by half of his own divided current from his own end of the line.

The puzzle to most people is: How can the signals pass each other in different directions on the same wire? But the signals do not have to pass each other. In effect, they pass; but in fact, it is like going round a circle--the earth forming half. A sends his message over the line to B. B sends his message to A through the earth and up A's ground-wire. The operative who is sending with positive pole to line pushes his current through--so to speak--while the operative who is sending with the negative pole to line pulls more current in the same direction through the line whenever he closes his key.

This may not be a strictly scientific statement; but, as long as we speak of a "current" flowing from positive to negative poles (which is the invariable course electricity takes), it is the way to look at the matter understandingly.

The short "resistance-coil" at each end, fortified by a "condenser" made of many leaves of isolated tin-foil, to give it capacity, offers precisely the same resistance to the current as the 500 miles of wire line; so that the twin currents that run around the differential magnet exactly neutralize each other and make no effect in the office the message starts from; while one of them takes to the earth, and the other to the line to carry the message.

This condenser is necessary, because the short resistance-coil affects the current immediately, while the long line with its greater amount of metal does not give the same amount of resistance till it is filled from end to end, which requires a fraction of a second. During this time, however, more current is passing through the differential coil connected with the line than through the short resistance-coil; and the unequal flow causes the relay armature to jump, or "kick." The condenser, with the many leaves of tin-foil, supplies the greater metal surface to be traversed by the short line current, causes the flow to be equal in both circuits at all times, and thus cures the "kick." It is this quality of a condenser that enables us to give to an artificial line of any resistance all the qualities, including capacity, and exhibit all the phenomena of a real line of any length, and it was this quality that enabled Mr. Stearns to take the "kick" out of duplex transmission and thus change the whole system, which created a new era in telegraphy.

We have just spoken of the "capacity" of a circuit, and stated that it was determined by the mass of metal used. This capacity is measured by a standard of capacity that is arbitrary and consists of a condenser, constructed so that a given amount of surface of tin-foil may be plugged in or out. The practical unit of capacity is called the micro-farad, the real unit is the farad, and takes its name from Faraday.

But let us go back to multiple systems of transmission. There are many other systems of simultaneous transmission aside from the duplex, and all of them are classed under the general head of multiple telegraphy. First there is the quadruplex, that sends two messages each way simultaneously, making one wire do the work of four single wires--as they were used at first. The quadruplex is very extensively used by the Western Union Telegraph Company and others. It would be difficult to explain it in a popular article, so we will not attempt it. There is another form of multiple telegraph that was used on the Postal Telegraph line when it first started--which was invented and perfected by the writer--that can be more easily explained.

In 1874 I discovered a method of transmitting musical tones telegraphically, and the thing that set my mind in that direction was a domestic incident. It is a curious fact that most inventions have their beginnings in some incident or observation that comes within the experience of some one who is able to see and interpret the meaning of such incidents or observations. I do not mean to say that inventions are usually the result of a happy thought, or accident; the germ may be, but the germ has to have the right kind of soil to take root in and the right kind of culture afterward. It is a rare thing that an invention, either of commercial or scientific importance, ever comes to perfection without hard work--midnight oil and daylight toil; and it is rarely, if ever, that a discovery or an invention based upon a discovery does not have, sooner or later, a practical use, although we sometimes have to wait centuries to find it put. We had to wait forty-four years after the galvanic battery was discovered before it became a useful servant of man. It was fifty years or more after the discovery by Faraday of magneto-electricity before it found a useful application beyond that of a mere toy, but now it is one of the most useful servants we have, as shown in its wonderful development in electric lighting and electric railroads, to say nothing of its heating qualities and the useful purpose it serves in driving machinery. The interesting discoveries of Professor Crookes in passing a current of electricity through tubes of high vacua waited many years before they found a practical use in the X-ray, that promises to be of great service in medicine and surgery.

The transmission of musical harmonies telegraphically, while in itself of great scientific interest, was of no practical use, but it led to other inventions, of which it is the base, that are transcendently useful in every-day life. The transmission of harmonic sounds by electricity underlies the principle of the telephone. There is a vast difference, in principle, between the transmission of simple melody, which is a combination of musical tones transmitted successively--one tone following another--and the transmission of harmony, which involves the transmission of two or more tones simultaneously. The former can be transmitted by a make-and-break current. In the latter case one tone has to be superposed upon another and must be transmitted with a varying but a continuously closed current. I make a distinction between a closed circuit and a closed current. In the case of the arc-light the circuit is open (that is, broken), technically speaking, but the current is still flowing. The reason why the Reiss and other metallic contact telephone transmitters cannot successfully be used for telephone purposes is that metal points will not allow of sufficient separation of the transmitting points without breaking the current as well as the circuit. Carbon contacts admit of a much wider separation without actually stopping the flow of the current, which latter is a necessity for perfect telephonic transmission, and it was the use of carbon that made that form of transmitter a success.

There are other forms, or at least one other form that does not depend upon the length of the voltaic arc formed when the electrodes are separated. Of this we will speak another time. Now let us go back to the domestic incident referred to above.

One evening in the winter of 1873-4 I came home from my laboratory work and went into the bathroom to make my toilet for dinner. I found my nephew, Mr. Charles S. Sheppard, together with some of his playmates, taking electrical "shocks" from a little medical induction-coil that I heard humming in the closet. He had one terminal of the coil connected to the zinc lining of the bathtub--which was dry at that time--while he held the other in his left hand, and with his right was taking shocks from the lining of the tub by rubbing his hand against the zinc. I noticed that each time he made contact with the tub, as he rubbed it for a short distance, a peculiar sound was emitted from under his hand, not unlike the sound made by the electrotome that was vibrating in the closet. My interest was immediately aroused, and I took the electrode out of his hand and for some time experimented with it, going to the cupboard from time to time to change the rate of vibration of the electrotome, and thus change the quality of the sound. I noticed that the sound or tone under my hand, if it could be so called, changed with each change of the rate of vibration. The thing that most interested me was that the peculiar characteristics of the noise were reproduced. In those few minutes I laid out work enough for years of experiment, and as a result I was late to dinner.

This discovery opened up to my mind the possibility of three things--the transmission of music and of speech or articulate words through a telegraph-wire, and the transmission of a number of messages over a single wire. I constructed a keyboard consisting of one octave and made a set of reeds tuned to the notes of the scale, and then when some one would play a melody I could reproduce it in two ways: One by placing my body in the circuit and rubbing a metal plate--it might be the bottom of a tin pan, a joint of stovepipe or otherwise--anything that was metal and would vibrate would give the effect. Another way was to connect an electromagnet (having a diaphragm or reed across its poles) in the circuit at the receiving-end and mount it on some kind of a soundboard. I made a great number of different kinds of receivers that were capable of receiving either musical or articulate sounds, as has many times been proven by experiment. I carried two sets of experiments along together; the one looking toward a system of multiple telegraphy and the other the transmission of articulate speech. Let us first look into the multiple telegraph and take the other up under the head of the telephone.

When the electrical keyboard was completed I found that I could transmit not only a melody but a harmony; that more than one tone could be transmitted simultaneously. This discovery opened up a long series of experiments with the view of sending a number of messages simultaneously by means of musical tones differing in pitch. I had already demonstrated that several tones could be transmitted at once, but they would speak all alike (with the same loudness) on the receiving-instrument. I now went to work on an instrument that responded for one note only and succeeded beyond my expectations. I made three different kinds of receiving-instruments. The first was a steel strap about eight inches long by three-eighths wide. This strap was mounted in an iron frame in front of an electromagnet. A thumbscrew enabled me to stretch the strap till it would vibrate at the required pitch. If, for instance, the sending-reed vibrated at the rate of 100 times per second and the strap of the receiver was stretched to a tension that would give 100 vibrations per second when plucked, it would then respond to the vibrations of the sending-reed but not to those of another reed of a different rate of vibration. If we take mounted tuning-forks tuned in pairs of different pitches, say four pairs, so that each fork has a mate that is in exact accord with it, and place them all in the same room, and sound one of them for a few seconds and then stop it, upon examining the other forks you will find all of them quiet except the mate of the one that was sounded. This one will be sounding. If we now sound four of the forks and then stop them the other four will be sounding from sympathy because the mate of each one of them has been sounded. If only two forks differing in pitch are sounded only two of the others will sound in sympathy. In the first case only one set of sound-waves were set up in the air, and the fork that found itself in accord with this set responded. When four forks differing in pitch were sounded there were four sets of tone-waves superposed upon each other existing in the air, so that each of the remaining forks found a set of waves in sympathy with its own natural rate of vibration and so responded.

Now apply this principle to the harmonic telegraph and you can understand its operation. At the transmitting-end of a line of wire there are a certain number of forks or reeds kept vibrating continuously. These reeds each have a fixed rate of vibration and bear a harmonic relation to each other so as not to have sound-interference or "beats." At the receiving-end of the line there are as many electromagnets as there are transmitting-reeds, and each magnet has a reed or strap in front of it tuned to some one of the transmitting-reeds, so that each transmitting-reed has a mate in exact harmony with it at the receiving-end of the line. Keys are so arranged at the transmitting-end as to throw the tones corresponding to them to line when depressed. In other words, when the key belonging to battery B and vibrator 1 is depressed (see Fig. 4) the effect is to send electrical pulsations through the line corresponding in rate per second to that of the vibrator. The same is true of battery B?and vibrator 2. During the time any key is depressed--we will say of tone No. 1--this tone will be transmitted through the line and be reproduced by its mate--the one tuned in accord with it--at the receiving-station. By a succession of long and short tones representing the Morse code a message can be sent. Numbers two, three and four might be sending at the same time, but they would not interfere with number one or with each other. In 1876-7 the writer succeeded in sending eight simultaneous messages between New York and Philadelphia by the harmonic method.

There were two ways of reading by the harmonic method. One was by the long and short tone-sounds and the other by the ordinary sounder.

The vibration of the receiving-reed was made to open and close a local circuit like a common Morse relay and thus operate the sounder. It is useless to try to send a message if the sender and receiver are out of tune with each other in this system.

What is true in science is true in life. If we are out of tune with our surroundings we only beat the air, and our efforts are in vain. We get no sympathetic response.

CHAPTER XIV.

WAY DUPLEX SYSTEM.

A novel form of double transmission was invented by the writer soon after the completion of the harmonic system, and was an outgrowth of it. It is still in use on some of the railroad-lines. An ordinary railroad telegraph-line has an instrument in circuit in every office along the road, chiefly for purposes of train-dispatching. As we have heretofore explained, whenever any one office is sending, the dispatch is heard in all of the offices. The "Way duplex" system permits of the use of the line for through business simultaneously with the operation of the local offices. That is to say, any station along the line may be telegraphing with any other station by the ordinary Morse method, and at the same time messages may be passing back and forth between the two end offices.

This is accomplished by the following method: At each end of the line there is a tuned reed, such as we have described in our last chapter, that is kept constantly in vibration by a local battery during working hours. This vibrator is so arranged in relation to the battery that whenever the key belonging to it is depressed the current all through the line is rendered vibratory. There is also in circuit at each end of the line a harmonic relay, that is tuned in accord with the vibrating reed of the sender. If either key belonging to this part of the system is opened, as in the act of sending a message, these harmonic relays, being tuned in sympathy with the sending-vibrator, will respond, thus sending Morse characters made up of a tone broken into dots and dashes. This tone can be read directly from the relay, or, as is usually the case, it causes the sounder to operate in the common way.

You will at once inquire why the ordinary Morse instruments in the local offices are not affected by these vibratory signals, and also why the harmonic instruments at the end office are not affected by the working of the local offices. The local office does not open the circuit entirely, but simply cuts out a resistance by the operation of the special harmonic key. When a resistance is thrown into an electric circuit it weakens the current in proportion to the amount of resistance interposed. You will see that there is some current still left in the line when the key is open, but the spring of the relay at the local office is so adjusted as to pull the armatures away from the magnets whenever the current is weakened by throwing in the resistance, so that by this means an ordinary Morse telegraphic relay may be worked without ever entirely opening the circuit. In the Way duplex system there is a resistance at each station that is cut in and out by the operation of its key, which causes all the instruments in the line to work simultaneously except the two harmonic relays located one at each end of the line. These will not respond to anything but the vibratory signal.

In order to prevent the Morse relays at the local offices from responding to the vibratory current a condenser is connected around them. This condenser serves two purposes: It enables the short impulses of the vibrating current to pass around the relays without having to be resisted by the coils of the magnets, and between the pulsations each condenser will discharge through the relay at the local offices, and thus fill in the gap between the pulsations, producing the effect on the relay of a steady current. When a line is thus equipped it may be treated in every respect as two separate wires, one of them doing way business and the other through business. It is a curious blending of science and mechanism.

Another interesting application was made of the system of transmission by musical tones--by Edison, some years ago. We refer to the transmission of messages to and from a moving railroad-train with the head office at the end of the line. In this case the message was transmitted a part of the distance through the air;--another instance of wireless telegraphy. The operation was as follows: One of the wires strung on the poles nearest to the track was fitted up with a vibrator and key at the end of the line similar to that of the Way duplex just described. In one of the cars was another battery, key and vibrator, and as only one tone was used, no tone-selecting device or harmonic relay was needed, but instead an ordinary receiving-telephone was used to read the long and short sounds sent over the lines. One end of the battery in the car was connected through the wheels to the earth, while the other end was connected to the metal roof of the car. Being thus equipped, we will suppose our train to be out on the road forty or fifty miles from either end of the line, moving at the rate of forty miles an hour. The operator at Chicago, say, wishes to send a message to the moving train; he operates his key in the ordinary manner, which makes the current on the line vibratory during the time the key is depressed. These electrical vibrations cause magnetic vibrations, or ether-waves, to radiate in every direction from the wire, at right angles to the direction of the current, like rays of light. When they strike the roof of the car they create electrical impulses in the metal by induction (described in Chap. VI). These impulses pass through a telephone located in the car to the ground. A Morse operator listening, with the telephone to his ear, will hear the message through the medium of a musical tone chopped up into the Morse code. In like manner the operator in the car may transmit a message to the roof of the car and thence through the air to the wire, which will be heard, by any one listening, in a telephone which is connected in that circuit,--and, as a matter of fact, it will be heard from any wire that may be strung on any of the poles on either side of the road.

Some years ago an experiment of this kind was made on one of the roads between Milwaukee and Chicago.

What wonderful things can be done with electricity! As a servant of man it is reliable and accurate--seeming almost to have the qualities of docility--when under intelligent direction, that is in accord with the laws of nature; but under other conditions it changes from the willing servant to a hard master, hesitating not to destroy life or property without regard to persons or things.

CHAPTER XV.

TELEPHONY.

In the foregoing chapters I have described the method of transmitting musical tones telegraphically and its applications to multiple telegraphy, as well as to a mode of communicating with a moving railroad-train. As I stated in a former chapter, after discovering a method of transmitting harmony as well as melody, I had in mind two lines of development, one in the direction of multiple telegraphy, and the other that of the transmission of articulate speech. I will not attempt to give the names of all the people who have contributed to the development of the telephone (as this alone would fill a volume) but only describe my own share in the work--leaving history to give each one due credit for his part. While I do not intend, here, to enter into any controversy regarding the priority of the invention of the telephone, I wish to say that from the time I began my researches, in the winter of 1873-4, until some time after I had filed my specification for a speaking or articulating telephone, in the winter of 1875-76, I had no idea that any one else had done or was doing anything in this direction. I wish to say further that if I had filed my description of a telephone as an application for a patent instead of as a caveat, and had prosecuted it to a patent, without changing a word in the specification as it stands to-day, I should have been awarded the priority of invention by the courts. I am borne out in this assertion by the highest legal authority. In law, a caveat (Latin word, meaning "Let him beware") is a warning to other inventors, to protect an incomplete invention; whereas in fact the invention to be protected may be complete. An application for a patent is presumed by the law to be for a completed invention; but it may be, and very often is, incomplete. It would often make a very great difference if decisions were rendered according to the facts in the case rather than according to rules of law and practice, that sometimes work great injustice to individuals.

As has been said in another chapter, in the summer of 1874 I went to Europe in the interest of the telephone, taking my apparatus, as then developed, with me. I came home early in the fall and resumed my experimental work. Many interesting as well as amusing things occurred during these experiments.

I remember that in the fall or early winter of 1874 I was in Milwaukee with my apparatus carrying on some experiments on a wire between Milwaukee and Chicago. I had my musical transmitter along, and one evening, for the entertainment of some friends at the Newhall House, a wire was stretched across the street from the telegraph office into one of the rooms of the hotel. A great number of tunes were played at the telegraph-office by Mr. Goodridge, who was my assistant at that time, which were transmitted across the street, as before stated. In those days it was a common practice in telegraphy to use one battery for a great number of lines. For instance, starting with one ground-wire which connected with, say, the negative pole of the battery, from the positive pole two, three or a half-dozen lines might be connected, running in various directions, connecting with the ground at the further end, thus completing their circuits. For use in transmitting tones across the street that evening we connected our line-wire on to the telegraph company's battery, which consisted of 100 or more cells, and which had four or five more lines radiating from the end of the battery to different parts of Wisconsin. Our line was tapped on to the battery (without changing any of its connections) twenty cells from the ground-wire. In transmitting, each vibration would momentarily shut off these twenty cells from the lines that were connected with the whole battery. The effect of this (an effect that we did not anticipate at the time) was to send a vibratory current out on all the lines that were connected with that single battery as well as across the street. A great many familiar tunes were played during the course of an hour or two which, unconsciously for us, were creating great consternation throughout the State of Wisconsin, in many of the offices through which these various lines passed.

Next morning reports and inquiries began to come in from various towns and cities west, northwest and north, giving details of the phenomena that were noticed on the instruments located in the various offices along the lines. They reported their relays as singing tunes; one party said he thought the instruments were holding a prayer-meeting from the fact that they seemed to be singing hymn-tunes for quite a while, but this notion was finally dissipated, because they grew hilarious and sang "Yankee Doodle."

One operator, up in the pine woods of northern Wisconsin, did not seem to take the cheerful view of it that some of the others did. He was sitting alone in the telegraph-office that evening when he thought he heard the notes of a bugle in the distance; he got up and went to the door to listen, but could hear nothing; but on coming back into the room he heard the same bugle notes very faintly. He was inclined to be somewhat superstitious and grew very nervous; finally, on looking around, he located the sound in his relay, but this did not help matters with him. With superstitious awe he listened to the instrument for a few moments, while it gave out the solemn tones of "Old Hundred," then it suddenly jumped into a hilarious rendering of "Yankee Doodle." This was too much for our nervous friend, and hastily putting on his overcoat, he left the office for the night.

On another occasion, when I was giving a lecture in one of the cities outside of Chicago, where exhibitions of music transmitted from Chicago were given, one of the operators along the line was very much astonished by his switchboard suddenly becoming musical. Orders had been given for the instruments in all the local offices to be cut out of the particular line that I was using. Hence the instrument in this particular office was not in the circuit through which the tunes were being transmitted. The wire, however, ran through his switchboard, and owing probably to a loose connection, or an induced effect, there was a spark that leaped across a short space at each electrical pulsation that passed through the line, thus reproducing the notes of the various tunes played.

You will remember in one of the chapters on sound (Volume II.), it is stated that a musical tone is made up of a succession of sounds repeated at equal intervals, and that the pitch of the tone is determined by the number of sound-impulses per second. Applying this law to the sparks, you will be able to see how the switchboard played tunes for the operator.

In the foregoing experiments in transmitting musical tones telegraphically, I used a great many different varieties of receivers. Some of them were designed with metal diaphragms mounted over single electromagnets, not unlike the receiver of an ordinary telephone. These instruments would both transmit and receive articulate speech when placed in circuit with the right amount of battery to furnish the necessary magnetism. However, they were not used in that way at the time they were first made--in 1874. These I called common receivers, as they were designed to reproduce all tones equally well. I designed and constructed another form of receiver, based somewhat upon the theory of the harmonic telegraph.

This consisted of an electromagnet of considerable size, mounted upon a wooden rod about ten feet long. Mounted upon this rod were also resonating boxes or tubes made of wood of the right size to have their air-cavities correspond with the various pitches of the transmitting-reeds, so that each tone would be re-enforced by some one of these air-cavities, thus giving a louder and more resonant effect to the musical notes.

Here were two types of receiver, one that would receive one sound as well as another, but none of them so loud, while the other was constructed on the principle of selection and re-enforcement, so that a particular note would be sounded by the box having a cavity corresponding to the pitch of the tone, and was much louder and of much better quality than I could get from the diaphragm receiver. One of these receivers pointed to the harmonic telegraph and the other to the speaking telephone. I knew that I had a receiver that would reproduce articulate speech or anything else that could be transmitted.

My first conceptions of an articulate speech-transmitter were somewhat complicated. I conceived of a funnel made of thin metal having a great number of little riders, insulated from the funnel at one end and resting lightly in contact with the funnel at the other end. These riders were to be made of all sizes and weights so as to be responsive to all rates of vibration. In the light of the present day we know that such an arrangement would have transmitted articulate speech, but perhaps not so well as a single point would do when properly adjusted. My mind clung to this idea till in the fall of 1875, when an observation I made upon the street changed the whole course of my thinking and solved the problem. The incident I refer to took place in Milwaukee, where I was then experimenting. One day while out on an errand I noticed two boys with fruit-cans in their hands having a thread attached to the center of the bottom of each can and stretched across the street, perhaps 100 feet apart. They were talking to each other, the one holding his mouth to his can and the other his ear. At that time I had not heard of this "lovers' telegraph," although it was old. It is said to have been used in China 2000 years ago.

The two boys seemed to be conversing in a low tone with each other and my interest was immediately aroused. I took the can out of one of the boy's hands (rather rudely as I remember it now), and putting my ear to the mouth of it I could hear the voice of the boy across the street. I conversed with him a moment, then noticed how the cord was connected at the bottom of the two cans, when, suddenly, the problem of electrical speech-transmission was solved in my mind. I did not have an opportunity immediately to construct an instrument, as I had a partner who was furnishing money for the development of the harmonic telegraph and would not listen to any collateral experiments. I remember sitting down by this partner one day and telling him what I could do in the way of transmitting speech through a wire. I told him I thought it would be very valuable if worked out. He gave me a look that I shall never forget, but he did not say a word. The look conveyed more meaning than all the words he could have said, and I did not dare broach the subject again.

However, as soon as I found opportunity, without saying a word to anybody except my patent lawyer, I filed a description, accompanied by drawings, of a speaking telephone which stands in history to-day as the first complete description on record of the operation of the speaking telephone. It described an apparatus which, when constructed, worked as described, and it is a matter of history that the first articulate speech electrically transmitted in this country was by a transmitter constructed on the principle described, and almost identically after the drawings in my caveat. While the transmitter described in this caveat was not the best form, it would transmit speech, and it contained the foundation principle of all the telephone transmitters in use to-day.

There are two methods of transmitting speech. One is known as the magneto method and the other that of varying the resistance of the circuit. My first transmitter was devised on the latter principle.

I append to this extracts from my specification filed Feb. 14, 1876:

To All Whom It May Concern:--Be it known that I, Elisha Gray of Chicago, in the County of Cook and State of Illinois, have invented a new art of transmitting vocal sounds telegraphically, of which the following is a specification: It is the object of my invention to transmit the tones of the human voice through a telegraphic circuit, and reproduce them at the receiving-end of the line, so that actual conversations can be carried on by persons at long distances apart. I have invented and patented methods of transmitting musical impressions or sounds telegraphically, and my present invention is based upon a modification of the principle of said invention, which is set forth and described in letters patent of the United States, granted to me July 27, 1875, respectively numbered 166,095 and 166,096, and also in an application for letters patent of the United States, filed by me, Feb. 23, 1875. * * * My present belief is that the most effective method of providing an apparatus capable of responding to the various tones of the human voice is a tympanum, drum, or diaphragm, stretched across one end of the chamber, carrying an apparatus for producing fluctuations in the potential of the electric circuit and consequently varying in its power. * * * The vibrations thus imparted are transmitted through an electric circuit to the receiving-station, in which circuit is included an electromagnet of ordinary construction, acting upon a diaphragm to which is attached a piece of soft iron, and which diaphragm is stretched across a receiving vocalizing chamber C, somewhat similar to the corresponding vocalizing chamber A.

The diaphragm at the receiving-end of the line is thus thrown into vibrations corresponding with those at the transmitting-end, and audible sounds or words are produced.

The obvious practical application of my improvement will be to enable persons at a distance to converse with each other through a telegraphic circuit, just as they now do in each other's presence, or through a speaking-tube.

I claim as my invention the art of transmitting vocal sounds or conversations telegraphically through an electric circuit.

This specification was accompanied by cuts of the transmitter and receiver connected by a line-wire and showing one person talking to the transmitter and another listening at the receiver. These cuts may be seen in various books on the subject of telephony.

CHAPTER XVI.

HOW THE TELEPHONE TALKS.

Everybody knows what the telephone is because it is in almost every man's house. But while everybody knows what it is, there are very few (comparatively speaking) that know how it works. If you remember what has been said about sound and electromagnetism it will not be hard to understand.

When any one utters a spoken word the air is thrown into shivers or vibrations of a peculiar form, and every different sound has a different form.

Therefore, every articulate word differs from every other word, not only as a shape in the air, but as a sensation in the brain, where the air-vibrations have been conducted through the organ of hearing; otherwise we could not distinguish between one word and another. Every different word produces a different sensation because there is a physical difference, as a shape or motion, in the air where it is uttered. If one word contains 1000 simultaneous air-motions and another 1500 you can see that there is a physical or mechanical difference in the air.

The construction of the simplest form of telephone is as follows: Take a piece of iron rod one-half or three-quarters of an inch long and one-quarter inch thick, and after putting a spool-head on each end to hold the wire in place wind it full of fine insulated copper wire; fasten the end of this spool to the end of a straight-bar permanent magnet. Then put the whole into a suitable frame, and mount a thin circular diaphragm (membrane or plate) of iron or steel, held by its edges, so that the free end of the spool will come near to but not touch the center of the diaphragm. This diaphragm must be held rigidly at the edges.

Now if the two ends of the insulated copper wires are brought out to suitable binding-screws the instrument is done.

The permanent steel magnet serves a double purpose. When the telephone was first used commercially, the instrument now used as a receiver was also used as a transmitter. As a transmitter it is a dynamo-electric machine. Every time the iron diaphragm is moved in the magnetic field of the pole of the permanent magnet, which in this case is the free end of the spool (the iron of the spool being magnetic by contact with the permanent magnet), there is a current set up in the wire wound on the spool; a short impulse, lasting only as long as the movement lasts. The intensity of the impulse will depend upon the amplitude and quickness of the movement of the diaphragm. If there is a long movement there will be a strong current and vice versa. If a sound is uttered, and even if the multitude of sounds that are required to form a word, be spoken to the diaphragm, the latter partakes in kind of the air-motions that strike it. It swings or vibrates in the air, and if it is a perfect diaphragm it moves exactly as the air does, both as to amplitude and complexity of movement. You will remember that in the chapter on sound-quality (Vol. II) it was said that there were hundreds and sometimes thousands of superposed motions in the tones of some voices that gave them the element we call quality.

All these complex motions are communicated by the air to the diaphragm, and the diaphragm sets up electric currents in the wire wound on the spool, corresponding exactly in number and form, so that the current is molded exactly as the air-waves are. Now, if we connect another telephone in the circuit, and talk to one of them, the diaphragm of the other will be vibrated by the electric current sent, and caused to move in sympathy with it and make exactly the same motions relatively, both as to number and amplitude.

It will be plain that if the receiving diaphragm is making the same motions as the transmitting diaphragm, it will put the air in the same kind of motion that the air is in at the transmitting end, and will produce the same sensation when sensed by the brain through the ear. If the air-motion is that of any spoken word it will be the same at both ends of the line, except that it will not be so intense at the receiving-end; it is the same relatively. And this is how the telephone talks.

I have said that the permanent magnet had two functions. In the case of the transmitter it is the medium through which mechanical is converted into electrical energy. It corresponds to the field-magnet of the dynamo, while the diaphragm corresponds to the revolving armature, and the voice is the steam-engine that drives it. In the second place, it puts a tension on the diaphragm and also puts the molecules of the iron core of the magnet in a state of tension or magnetic strain, and in that condition both the molecules and the diaphragm are much more sensitive to the electric impulses sent over the wire from the transmitter. This fact was experimented upon by the writer as far back as 1879 and published in the Journal of the American Electrical Society. At the present day this form of telephone is used only as a receiver.

Transmitters have been made in a variety of forms, but there are only two generic methods of transmission. One is the magneto method--the one we have described--and the other is effected by varying the resistance of a battery current. The former will work without a battery, as the voice acting on the wire around the magnet through the diaphragm creates the current; in the latter the current is created by the battery but molded by the voice. In the latter method the current passes through carbon contacts that are moved by the diaphragm. Carbon is the best substance, because it will bear a wider separation of contact without actually breaking the current. When carbon points are separated that have an electric current passing through them, there is an arc formed on the same principle as the electric arc-light.

Great improvements in details have been made in the telephone since its first use, but no new principles have been discovered as applied to transmission.

We have spoken in another place regarding the various claimants to the invention of the telephone, but here is one that has been overlooked. A young man from the country was in a telegraph-office at one time and was left alone while the operator went to dinner. Suddenly the sounder started up and rattled away at such a rate that the countryman thought something should be done. He leaned down close to the instrument and shouted as loudly as possible these words: "The operator has gone to dinner." From what we know now of the operation of the telephone I have no doubt but that he transmitted his voice to some extent over the wire. This young man's claims have never been put forward before, and we are doing him tardy justice. But his claim is quite as good as many others set forth by people who think they invent, whenever it occurs to them that something new might possibly be done, if only somebody would do it. And when that somebody does do it they lay claim to it.

In the early days of the telephone it was not supposed that a vocal message could be transmitted to a very great distance. However, as time went on and experiments were multiplied the distance to which one could converse with another through a wire kept on increasing.

In these days, as every one knows, it is a daily occurrence that business men converse with each other, telephonically, for a distance of 1000 miles or more; in fact, it is possible to transmit the voice through a single circuit about as great a distance as it is possible to practically telegraph. This leads us to speak of another telegraphic apparatus which we have not heretofore mentioned, and that is the telegraphic repeater. It is a common notion that messages are sent through a single circuit across the continent, but this is not the case, although the circuits are very much longer than they were some years ago. The repeater is an instrument that repeats a message automatically from one circuit to another. For instance, if Chicago is sending a message to New York through two circuits, the division being in Buffalo, the repeater will be located at Buffalo and under the control of both the operator at Chicago and the operator in New York. When Chicago is sending, one part of the repeater works in unison with the Chicago key and is the key to the New York circuit, which begins at Buffalo. When New York is sending the other part of the repeater operates, which becomes a key which repeats the message to the Chicago line. In this way the practical result is the same as though the circuit were complete from New York to Chicago. At the present day some of the copper wires and perhaps some of the larger iron wires are used direct from Chicago to New York without repetition, but all messages between New York and San Francisco are automatically repeated at least twice and under certain conditions of weather oftener. I can remember that in wet weather in the old days, with such wires as they had then (being No. 9 iron with bad joints, which gave the circuit a high resistance) that these repeaters would be inserted at Toledo, Cleveland, Buffalo and Albany in order to work from Chicago to New York. Under such conditions the transmission would necessarily be slow, because an armature time will be lost at each repeater. Regarding each repeater as a key, when Chicago depresses his key the armature of the next repeater must act, and then the next successively, and all of this takes time, although only a small fraction of a second.

The repeater was a very delicate instrument and had to be handled by a skilled operator. Every wire must be in its place or the instrument would fail to operate. I remember on one occasion in Cleveland that along in the middle of the night the repeater failed to work. The operator knew nothing of the principle of its operation, so that when it failed he had to appeal to some of his superiors.

At this time there was no one in the office who knew how to adjust it, so they had to send up to the house of the superintendent and arouse him from his sleep and bring him down to the office. He looked under the table and found that one of the wires had loosened from its binding-post and was hanging down. He said immediately, "Here's the trouble; I should think you could have seen it yourself." The operator replied, "I did see that, but I didn't think one wire would make any difference." He learned the lesson that all electricians have had to learn--that even one wire makes all the difference in the world. But this operator was no worse in that respect than some of his superiors. One of the heads of the Cleveland office at one time in the early days wanted to give some directions to the office at Buffalo. He told the operator at the key to tell Buffalo so and so, when the operator replied: "I can't do it; Buffalo has his key open." The official immediately said with severity: "Tell him to close it." He forgot that it would be as difficult for him to tell him to close it, as it would have been to have sent the original message.

But let us go back to the telephone. While it is possible to send a message from New York to San Francisco by telegraph, it is not possible to telephone that distance, because as yet no one has been able to devise a repeater that will transfer spoken words from one line to another satisfactorily. But unless the printer and publisher bestir themselves some one may accomplish the feat before this little book reaches the reader. If this proves to be true, let the writer be the first to congratulate the successful inventor.

CHAPTER XVII.

SUBMARINE CABLES.

The first attempts at transmitting messages through wires laid in water were made about 1839. These early experiments were not very successful, because the art of wire-insulation had not attained any degree of perfection at that time. It was not until gutta-percha began to be used as an insulator for submarine lines that any substantial progress was made.

The first line, so history states, that was successfully laid and operated was across the Hudson River in 1848. This line was constructed for the use of the Magnetic Telegraph Company.

In the following year experiments with gutta-percha insulation were successfully made, and about 1850 a cable was laid across the English Channel between Dover and Calais (twenty-seven miles), consisting of a single strand of wire having a covering of gutta-percha. The insulation was destroyed in a day or two, which demonstrated the fact that all submarine cables must be protected by some kind of armor. In 1851 another cable was laid between these two points, containing four conductors insulated with gutta-percha, and over all was an armor of iron wire. Twenty-one years later this cable was still working, and for all we know is working now. After this successful demonstration other cables were laid for longer distances.

These short-line cables served to demonstrate the relative value of different material for insulating purposes under water, and it has been found that gutta-percha possesses qualities superior to almost every other material as an insulator for submarine cables, although there are many better materials for air-line insulation. Gutta-percha when exposed to air becomes hardened and will crack, but under water it seems to be practically indestructible.

Ocean telegraphy really dates from the laying of the first successful Atlantic cable. There were many problems that had to be solved, which could be done only by the very expensive experiment of laying a cable across the Atlantic Ocean. In the first place a survey had to be made of the bottom of the ocean between the shores of America and Great Britain. The most available route was discovered by Lieutenant Maury of the United States Navy, who made a series of deep-sea soundings, and discovered that, from Newfoundland to the west coast of Ireland the bottom of the ocean was comparatively even, but gradually deepening toward the coast of Ireland until it reached a depth of 2000 fathoms. It was not so deep but that the cable could be laid on the bottom, nor so shallow as to be in danger of the waves, icebergs or large sea-animals.

The water below a certain depth is always still and not affected by winds or ocean currents. At many other points in crossing the ocean, high mountains and deep valleys are encountered, possessing all the topographical features of dry land--as the ocean bed is only a great submerged continent.

The beginning of the laying of the first Atlantic cable was on Aug. 7, 1857. On the morning of Aug. 7, 1858, a year later, after a series of mishaps and adverse circumstances that would have discouraged most men, the country was electrified by a dispatch from Cyrus W. Field of New York (to whom the final success of the Atlantic cable is mainly due), that the cable had been successfully laid and worked. But this cable worked only from the 10th of August to the 1st of September, having sent in that time 271 messages. The insulation became impaired at some point, when an attempt to force the current through by means of a large battery only increased the difficulty.

The failure of this first cable served to teach manufacturers and engineers how to construct cables with reference to the conditions under which they are to be used. It was found that in the deep sea a much smaller and less expensive cable could be used than would answer at the shore ends, where the water is shallow. The shore ends of an ocean cable are made very large, as compared to the deep-sea portions, so as to resist the effect of the waves and other interfering obstacles. It was further learned that the most successful mode of transmitting signals through the cable was with a small battery of low voltage, and by the use of very delicate instruments for receiving the messages. It is not possible to employ such instruments on cables as are used on land-lines, while it would not be a difficult feat to transmit even twice the distance over land-lines strung on poles, using the ordinary Morse telegraph.

The water of the ocean is a conductor, as well as the heavy armor that surrounds the insulation of the cable. When a current is transmitted through the conducting wires, in the center of the cable, they set up a countercharge in the armor and the water above it, somewhat as an electrified cloud will induce a charge in the earth under it, of an opposite nature. This countercharge, being so close to the conducting wire, has a retarding effect upon the current transmitted through it. An ordinary land-line that is strung on poles that are high up from the ground has this effect reduced to a minimum, but the greater the number of wires clustered together on the same poles the more difficult it becomes to send rapid signals through any one of them.

The instrument used for receiving cable messages was devised by Sir William Thompson, now Lord Kelvin. One form consists of a very short and delicate galvanometer-needle carrying a tiny mirror. This mirror is so related to a beam of light thrown upon it that it reflects it upon a graduated screen at some distance away, so that its motions are magnified many hundred times as it appears upon the screen. An operator sits in a dark room with his eye on the screen and his hand upon the key of an ordinary Morse instrument. He reads the signal at sight, and with his key transmits it to a sounder, which may be in another room, where it is read and copied by another operator. Another form of receiving-instrument carries, instead of the mirror, a delicate capillary glass tube that feeds ink from a reservoir, and by this means the movements of the needle are recorded on a moving strip of paper. The symbols (representing letters) are formed by combinations of zigzag lines.

This instrument is the syphon-recorder.

CHAPTER XVIII.

SHORT-LINE TELEGRAPHS.

Early in the history of the telegraph short lines began to be used for private purposes, and as the Morse code was familiar only to those who had studied it and were expert operators on commercial lines, some system had to be devised that any one with an ordinary English education could use; as the expense of employing two Morse operators would be too great for all ordinary business enterprises. These short lines are called private lines, and the instruments used upon them were called private-line telegraph-instruments. Of course they are now nearly all superseded by the telephone, but they are a part of history.

One of the earliest forms of short-line instruments was called the dial-telegraph. One of the first inventors, if not the first, of this form of instrument was Professor Wheatstone of England, who perfected a dial-telegraph-instrument about the year 1839. The receiving-end of this instrument consisted of a lettered dial-face, under which was clockwork mechanism and an escape-wheel controlled by an electromagnet. Each time the circuit was opened or closed the wheel would move forward one step, and each step represented one of the letters of the alphabet, so that the wheel, like the type-wheel of a printing telegraph, had fourteen teeth, each tooth representing two steps. As the reciprocating movement of the escapement had a pallet or check-piece on each side of the wheel, its movement was arrested twenty-eight times in each revolution. These twenty-eight steps correspond to the twenty-six letters of the alphabet, a dot and a space. On the shaft of the escape-wheel is fastened a hand or pointer, which revolves over a dial-face having the twenty-six letters of the alphabet, also a dot and space. The pointer was so adjusted that when the escape-wheel was arrested by one of the pallets it would stop over a letter, showing thus, letter by letter, the message which the sender was spelling out.

The transmitter consisted of a crank with a knob and a pointer on it, which was mounted over a dial that was lettered in the same way as the face of the receiving-instrument. A revolution of this crank would break and close the circuit twenty-eight times; that is to say, there were fourteen breaks and fourteen closes of the circuit. If now the transmitting-pointer and the receiving-pointer are unified so that they both start from the same point on the dial, and the transmitting-crank is rotated from left to right, the receiving-pointer will follow it up to the limit of its speed. In transmitting a message the sender would turn his crank, or pointer, to the first letter of the word he wished to transmit, making a short pause, and then move on to the next letter, and so on to the end of the message, making a short pause on each letter. The end of a word was indicated by turning the pointer to the space-mark on the dial. The receiving-operator would read by the pauses of the needle on the various letters. This was a system of reading by sight.

There have been many forms of this dial-telegraph worked out by different inventors at different times, and quite a number of them were used in the old days. It was a slow process of telegraphing, but it was suited to the age in which it flourished. One of the difficulties of a dial-telegraph consisted in the readiness with which the transmitter and receiver would get out of unison with each other; and when this happened of course a message is unintelligible, and you have to stop and unify again.

About 1869 the writer invented a dial-telegraph to obviate this difficulty. In this system a transmitter and receiver were combined in one instrument, and instead of a crank there were buttons arranged around the dial in a circle, one opposite each letter. When not in operation the pointers of both instruments at both stations stood at zero. In the act of transmitting the operator would depress the button opposite the letter he wished to indicate, when immediately the pointers of both instruments would start up and move automatically, step by step, until the pointer came in contact with the stem of the depressed button, when it would be arrested, and at the same time cut out the automatic transmitting-mechanism and cause both needles to remain stationary during the time the button was depressed. Upon releasing the button the pointers both fall back to zero at one leap.

The first private line equipped by this instrument was for Rockefeller, Andrews & Flagler, which was the firm name of the parties who afterward organized the Standard Oil Company. This line was built between their office on the public square in Cleveland and their works over on the Cuyahoga flats.

It seemed, however, to be the fate of the writer to make new inventions that would supersede the old ones before they were fairly brought into use. Very soon after the dial-telegraph began to be used, printing telegraph instruments for private-line purposes superseded them. About 1867 a printing instrument was devised for stock reporting, which in one of its forms is still in use. Soon after the invention of this form of printer a company was organized to operate not only these stock-reporting lines, but short lines for all sorts of private purposes. Following the invention of the stock-reporting instrument there were several adaptations made of the printing telegraph for private-line purposes. Among others the writer invented one known as "Gray's automatic printer," a cut and a description of which may be found on page 684 in "Electricity and Electric Telegraph," by George B. Prescott, published in 1877. This instrument was adopted by the Gold and Stock Telegraph Company as their standard private-line printer. It was first introduced in the year 1871, and at the time the telephone began to be used there were large numbers of these printers in operation in all of the leading cities and towns in the United States. While this has been superseded to a large extent by the telephone, there are still a few isolated cases where it is used.

Short lines have multiplied for all sorts of purposes, until to-day the money invested in them largely exceeds the amount invested in the regular commercial telegraphic enterprises.

The invention of the telephone created such a demand for short-line service that some scheme had to be devised not only to make room for the necessary wires, but to so cheapen the instruments as to bring them within reach of the ability of the ordinary man of business.

This problem has been solved (but not without many difficulties) by the inauguration of what is known as the "central station." By this system one party simply controls a single wire from his office or residence to the central station; here he can have his line connected with any other wire running into this same station, by calling the central operator and asking for the required number. It is useless to tell the public that very often this number is "busy," and here is the great drawback to the central-station system. This is especially true in large cities, where there are a great number of lines. The switchboards in large cities are necessarily very complicated affairs, and it requires a number of operators to answer the many calls that are constantly coming in. Each central-station operator presides over a certain section of the board, and as this section has to be related in a certain way to every other section, it is easy to see wherein arises the complication.

In large cities the central stations themselves have to be divided and located in different districts, being connected by a system of trunk lines.

CHAPTER XIX.

THE TELAUTOGRAPH.

So far we have described several methods of electrical communication at a distance, including the reading of letters and symbols at sight (as by the dial-telegraph and the Morse code embossed on a strip of paper); printed messages and messages received by means of arbitrary sounds, and culminating in the most wonderful of all, the electrical transmission of articulate speech.

None of these systems, however, are able to transmit a message that completely identifies the sender without confirmation in the form of an autograph letter by mail.

In 1893 there was exhibited in the electrical building at the World's Fair an instrument invented by the writer called the Telautograph. As the word implies, it is a system by which a man's own handwriting may be transmitted to a distance through a wire and reproduced in facsimile at the receiving-end. This instrument has been so often described in the public prints that we will not attempt to do it here, for the reason that it would be impossible without elaborate drawings and specifications. It is unnecessary to state that it differs in a fundamental way from other facsimile systems of telegraphy. Suffice it to say that as one writes his message in one city another pen in another city follows the transmitting-pen with perfect synchronism; it is as though a man were writing with a pen with two points widely separated, both moving at the same time and both making exactly the same motions. By this system a man may transact business with the same accuracy as by the United States mail, and with the same celerity as by the electric telegraph.

A broker may buy or sell with his own signature attached to the order, and do it as quickly as he could by any other method of telegraphing, and with absolute accuracy, secrecy and perfect identification.

In 1893, when this apparatus was first publicly exhibited, it operated by means of four wires between stations, and while the work it did was faultless, the use of four wires made it too expensive and too cumbersome for commercial purposes; so during all the years since then the endeavor has been to reduce the number of wires to two, when it would stand on an equality with the telephone in this respect. It is only lately that this improvement has been satisfactorily accomplished, and, for reasons above stated, no serious attempt has been made to introduce it as yet; but it has been used for a long enough time to demonstrate its practicability and commercial value. Companies have been organized both in Europe and America for the purpose of putting the telautograph into commercial use.

By means of a switch located in each subscriber's office the wires may be switched from a telephone to a telautograph, or vice versa, in a moment of time. By this arrangement a man may do all the preliminary work of a business transaction through the telephone, and when he is ready to put it into black and white switch in the telautograph and write it down. For ordinary exchange work this is undoubtedly the true way to use the telautograph, because one system of wires and one central-station system will answer for both modes of communication, and in this way an enormous saving can be made to the public. There is no question in the mind of any one who is familiar with the operation of both the telephone and telautograph but that some day they will both be used, either in the same or separate systems, as they each have distinctly separate fields of usefulness,--the telephone for desultory conversation, the telautograph for accurate business transactions. The question may arise in the minds of experts how the two systems can be worked in the same set of cables, and this leads us to discuss the phenomena of induction.

Every one who has listened at a telephone has heard a jumble of noises more or less pronounced, which is the effect of the working of other wires in proximity to those of the telephone. If, when a Morse telegraph instrument is in operation on one of a number of wires strung on the same poles, we should insert a telephone in any one of the wires that were strung on the same poles or on another set of poles even across the street, we could hear the working of this Morse wire in the telephone, more or less pronounced, according to the distance the wire is from the Morse circuit. This phenomenon is the result of induction, caused by magnetic ether-waves that are set up whenever a circuit is broken and closed, as explained in Chapter VI.

The telephone is perhaps the most sensitive of all instruments, and will detect electrical disturbances that are too feeble to be felt on almost any other instrument, hence the telephone is preyed upon by every other system of electrical transmission, and for this reason has to adopt means of self-protection. It has been found that the surest way to prevent interference in the telephone from neighboring wires is to use what is called a metallic circuit--that is to say, instead of running a single wire from point to point and grounding at each end, as in ordinary telegraph systems, the telephone circuit is completed by using a second wire instead of the earth.

As a complete defense against the effects of induced currents the wires should be exactly alike as to cross-section (or size) and resistance. They should be insulated and laid together with a slight twist. This latter is to cause the two wires so twisted to average always the same distance from any contiguous wire.

One factor in determining the intensity of an induced current is the distance the wire in which it flows is from the source of induction. A telephone put in circuit at the end of the two wires that are thus laid together will be practically free from the effects of induced currents that are set up by the working of contiguous wires--for this reason: Whenever a current is induced in one of the slack-twisted wires it is induced in both alike; the two impulses being of the same polarity meet in the telephone, where they kill each other. In order to have a perfect result we must have perfect conditions, which are never attained absolutely, but nearly enough for all practical purposes.

In the early days of telephony great difficulty was experienced in using a single wire grounded at each end in the ordinary way, if it ran near other wires that were in active use. As time passed on and the electric light and electric railroad came into operation these difficulties were immensely increased, till now in large cities the telephone companies are fast being driven to the double-wire system, which will soon become universal for telephonic purposes the world over, except perhaps in a few country places where there is freedom from other systems of electrical transmission. To successfully work the telephone and telautograph through the same cables, these protective devices against induction must be very carefully provided and maintained.

CHAPTER XX.

SOME CURIOSITIES.

Until within recent years it was never supposed that a sunbeam would ever laugh except in poetry. But the modern scientist has taken it out of the realm of poetry and put it into the prosy play of every-day life. The Radiophone, invented by A. G. Bell, is an instrument by which articulate or other sounds are transmitted through the medium of a ray of light. It has as yet no practical application and has never gone beyond the experimental stage, but as a bit of scientific information it is very interesting.

If we introduce into an electric circuit a piece of selenium, prepared in a certain way, its resistance as an electric conductor undergoes a radical change when a beam of sunlight is thrown upon it. For instance, a selenium cell, so called, that in the dark would measure 300 ohms resistance, would have only about 150 ohms when exposed to sunlight. This amount of variation in a short circuit of low resistance would produce a considerable change in the strength of a current passing through it from a battery of a given voltage.

If now we connect a selenium cell to one pole of a battery, and thence through a telephone and back to the other pole, we have completed an electric circuit, of which the selenium cell is a part, and any variation of resistance in this cell, if made suddenly, will be heard in the telephone. Let the diaphragm of a telephone transmitter have a very light, thin mirror on one side of it, and a beam of sunlight be thrown upon it and reflected from that on to the selenium cell, which may be some distance away. Then, if the diaphragm is thrown into vibration by an articulate word or other sound, the light-ray is also thrown into vibration, which causes a vibratory change of resistance in the selenium cell in sympathy with the light-vibrations; and this in turn throws the electric current into a sympathetic vibratory state which is heard in the telephone. So that if a person laughs or talks or sings to the diaphragm, the sunbeam laughs, talks and sings and tells its story to the electric current, which impresses itself upon the telephone as audible sounds--articulate or otherwise. Instead of the telephone, battery and selenium cell, a block of vulcanite or certain other substances may be used as a receiver; as a light-ray thrown into vibration has the power to produce sound or sympathetic vibration in certain substances.

Another curious application of the selenium cell has been attempted, but has scarcely gone beyond the domain of theory. This apparatus, if perfected, might be called a Telephote. It is an apparatus by which an illuminated picture at one end of a line of many wires is reproduced upon a screen at the other end. The light is not actually transmitted, but only its effects. Suppose a picture is laid off into small squares and there is a selenium cell corresponding to each square and for each selenium cell there is a wire that runs to a distant station in which circuit there is a battery. At the distant station there are little shutters, one for each wire, that are controlled by the electric current and so adjusted that when the cell at the transmitting-end is in the dark the shutter will be closed. Now if a strong light be thrown upon the picture at the transmitting-end, and each square of the picture reflects the light upon its corresponding selenium cell, the high lights of the picture will reflect stronger light than the shadows, and therefore the wires corresponding to the high-light squares will have a stronger current of electricity flowing through them, because the resistance of the circuit is less than the ones connected with the darker shadows. So that the degree of current-strength in the various wires will correspond to the intensity of light reflected by the different sections of the picture. The shutters are so adjusted that the amount of opening depends upon the strength of current. The shutters corresponding to the high lights of the picture will open the widest and throw the strongest light upon the screen, from a source of light that is placed behind the shutters. The shutters that open the least will be those that are operated upon by the shadows of the picture. Inasmuch as a picture thrown on a screen from a source of light is wholly made up of lights and shadows, the theory is that this apparatus perfectly constructed would transmit any picture to a distance, through telegraph-wires. It must not be understood that the rays of light are transmitted through the wires as sound-vibrations are. Light, per se, can be transmitted only through the luminiferous ether, as we have seen in the chapter on light in Volume II.

While we are talking about these curious methods of telegraphic transmission, I wish to refer to an apparatus constructed by the writer in 1874-5, for the purpose of receiving musical tones or compositions transmitted from a distance through a wire by electricity. (A cut of this apparatus is shown on page 875 of "Electricity and Electric Telegraph," by Prescott, issued in 1877.) It consists of a disk of metal rotated by a crank mounted on a suitable stand. The electric circuit passes through the disk to the hand of the operator in contact with it, thence running through the line-wire to the distant station. Now, if a tune is played at that station, upon an electrical key-board, as described in a previous chapter, and the disk rotated with the fingers in contact with it, the tune or other sounds will be reproduced at the ends of the fingers. After the telephone was invented and put into use I used this revolving disk as a receiver for speech as well as music, and by this means persons may carry on an oral conversation through the ends of their fingers. This apparatus has been confounded in the minds of some people with Edison's electromotograph. The phenomena of the electromotograph were produced by chemical effects, while that of the apparatus just described is electrostatic in its action. The electrostatic disk was made in the winter of 1873-4, while Edison's electrochemical discovery was made some time later.

CHAPTER XXI.

WIRELESS TELEGRAPHY.

Broadly speaking, "Wireless Telegraphy" is any method of transmitting intelligible signals to a distance without wires; and this includes the old Semaphore systems of visual signals, such as flags and long arms of wood by day, and lights by night; also the Heliograph (an apparatus for flashing sunlight), and Sound Signals, made either through the air or water. Electrical conduction, either through rarefied air or the earth, also comes under this heading.

The name "Wireless Telegraphy," however, is specifically applied to a system of signaling by means of ether-waves induced by electrical discharges of very high voltage. Ether-waves of a greater or less degree are always set up whenever there are sudden electrical disturbances, however slight. Ether-waves, electrically induced, are probably as old as the universe. When "there were thunders and lightnings" from the cloud that hovered over Mount Sinai in the time of Moses, ether-waves of great power were sent out through the camp of Israel. But the people of those days had no "coherer" or telephone or any other means of converting these waves into visual or audible signals. Thousands of years had to elapse before the intellect of man could grasp the meaning of these natural phenomena sufficiently to harness them and make them subservient to his will.

Many people have been powerfully "shocked"--some even killed--by the impact of ether-waves set up by powerful discharges of lightning between the clouds and the earth--when they were not in the direct path of the lightning-stroke.

The history of Electro-Wireless Telegraphy, like that of all inventions, is one of successive stages, and all the work was not done by one man. The one who gets the most credit is usually the one who puts on the finishing touches and brings it out before the public. He may have done much toward its development or he may have done but little.

In the year 1842 Morse transmitted a battery current through the water of a canal eighty feet wide so as to affect a galvanometer on the opposite side from the battery. This was wireless telegraphy by conduction through water.

In 1835 Joseph Henry produced an effect on a galvanometer by ether-waves through a distance of twenty feet by an arrangement of batteries and circuits like that shown in Fig. 1, Chapter VI. This was called induction, and is still so called when electrical effects are produced from one wire to another through the ether for short distances. All induction-coils and transformers (see Chapter XXIV) are operated by effects produced through the ether from the primary to the secondary coil--but through very short distances.

In 1880 Professor Trowbridge transmitted an electrical current through the earth for one mile so as to produce signals in a telephone. In 1881-2 Professor Dolbear used for a short distance (fifty feet) substantially the same arrangement as Marconi now uses, except that the former used a telephone as a receiver. He used an induction-coil having one end of the secondary wire connected with the earth, while the other was attached to a wire running up into the air. At the receiving-end a wire starting from the earth extended into the air, passing through a telephone, which acted as a receiver. In 1886 he used a kite to elevate the wire, through which electrical discharges of high voltage were made into the air to produce ether-waves--the receiver being 2000 feet away. Dolbear's experiments were public fourteen years ago, but at that time there was no interest in such matters, so that his work received little or no attention. In 1887 Dr. Hertz of Germany made some experiments in producing and detecting ether-waves, and he did a great deal to awaken an interest in the subject, so that others began investigations that have led to its present use as a means of telegraphing to a distance of many miles.

In 1891 Professor Branly of Paris invented the coherer. In 1894 it was improved by Lodge and by him used as a detector of ether-waves. In 1896, ten years after Dolbear had used it with the kite at the transmitting-end and telephone at the receiving-end, Marconi, an Italian, substituted the coherer of Branly for the telephone of Dolbear. This coherer is constructed and operated as follows:

It consists of a glass tube, of comparatively small diameter, loosely filled with metal filings of a certain grade. This body of metal-dust is made a part of a local battery circuit in which is placed an ordinary electric bell or telegraphic sounder. The resistance of this body of filings is so great that current enough will not pass through it to ring the bell or actuate the sounder until an ether-wave strikes it and the wire attached to it, when the metal particles are made to cohere to such an extent that the conductivity of the mass is greatly increased; so that a current of sufficient volume will now pass through the bell-magnet to ring it. Before the next signal comes the filings must be made to de-cohere; and to accomplish this a little "tapper," that works automatically between the signals, strikes the glass tube with a succession of light blows.

Briefly stated, the wireless system of Marconi, in its essentials, consists of a powerful induction-coil with one end of the secondary wire connected with the earth, while the other extends into the air a greater or less distance according to the distance it is desired to send signals. The greater the distance the higher the wire should extend into the air. At the receiving-end a wire of corresponding height is erected, also connected with the earth. In this wire--as a part of its circuit--is placed the coherer. In a local circuit that is connected to the upright wire in parallel with the coherer is placed a battery, a sounder, or a bell, that is rung when the filings cohere.

When an ether-wave is set up by a discharge of electricity into the air it strikes the perpendicular wire of the receiver, and that portion of the wave that strikes is converted into electricity, which is called an induced current. It is this current, as it discharges through the coherer to the earth, that causes the filings to unite so as to close the local circuit and operate the sounder. To send a message it is only necessary to make the discharges into the air, at the sending-end, correspond to the Morse alphabet.

While Marconi has done more than any other man to improve and popularize wireless telegraphy, history shows that he invented none of the essential elements so far as the system has been made public.

What he seems to have really done was to substitute the coherer of Branly and Lodge, with its adjuncts, for the telephone of Dolbear. There is no doubt but that Marconi has done much to improve and enlarge the capacity of the apparatus and to demonstrate to the world some of its possibilities. He has been an indefatigable worker and deserves great credit; but without the work of those who preceded him he could not have succeeded: the honors should be divided.

This system has been used at various times for reporting yacht-races, and between ships. It is said also to have been used to some extent in the South African War. There is much to be done yet, however, before it can be made entirely reliable for defensive work in time of war. As it is now, all an enemy would have to do to destroy its usefulness would be to set an ether-wave-producer to work automatically anywhere within the "sphere of influence" of the system--to speak diplomatically--when it would render unintelligible any message that should be sent. To make the system of the greatest value some sort of selective receiver must be invented that will select signals sent from a transmitter that is designed to work with it.

There is no doubt but that wireless telegraphy will some time play an important part in many spheres of usefulness.

There is another mode (already referred to) for transmitting signals electrically without wires through the earth instead of through the air, but in this case it is not through the medium of induction, but conduction. It has been explained in former chapters that earth-currents are constantly flowing from one point to another where the potentials are unequal. Sometimes these inequalities of potential are caused by heat and sometimes by electricity, as in the case of a thunder-storm. If a cloud is heavily charged with positive electricity, say, the earth underneath will have an equal charge of negative electricity. Let us illustrate it by the tides. As the moon passes over the ocean it attracts the water toward it and tends to pile up, as it were, at the nearest point between the earth and the moon. Suppose that (while the water is thus piled up at a point under the moon) we could suddenly suspend the attraction between the earth and the moon--the water would begin immediately to flow off by the force of gravitation until it had found a common level. Suppose in the place of the moon we have a cloud containing a static charge of positive electricity--it attracts a negative charge to a point on the earth nearest the cloud. If now a discharge takes place between the earth and cloud the potential between the two will suddenly become equalized and the static charge that was accumulated in the earth is released and it dissipates in every direction, seeking an equilibrium, following the analogy of the water; the difference being that in one case the movement is very slow, while in the other it is as "quick as lightning."

About eighteen years ago I had a short telephone-line between my house and that of one of my neighbors. This line was equipped with what was known in those days as magneto-transmitters, such as we have described in a previous chapter on the subject of telephony. When a line is equipped in this way no batteries are needed, as the voice generates the current, on the principle employed in the dynamo-electric machine. Often on summer evenings, when the sky appears to be cloudless, we can see faint flashes of lightning on the horizon, an appearance which is commonly called "heat-lightning." As a matter of fact, I do not suppose there is any such thing as heat-lightning, but what we see is the effect of very distant storm-clouds. Often at such times I have held the telephone receiver to my ear and could hear simultaneously with each flash a slight sound in the telephone. This effect could be produced in the earth by a simple discharge between two or more clouds, which would distribute the electrical discharge over a greater area. And because my line had connection with the earth it could have been disturbed electrically by conduction instead of induction; or it may have been the effect of ether-waves set up by the lightning discharges. There is no doubt in my mind but that both of these effects (ether-waves and conduction through earth) may be felt when a discharge takes place between a cloud and the earth.

If we could, by operating an ordinary telegraphic key, cause the lightning to discharge from cloud to earth, and some one was listening at a telephone in a circuit that was grounded at both ends 100 miles or more distant from the cloud, the man who controlled the discharges by the key could transmit the Morse code through the earth to the man who was listening at the telephone. Thousands of people might be listening at telephones in every direction from the transmitting-station, and they would all get the same message. If the receiving-station is near to the point where there is a heavy discharge from the clouds to the earth the earth-current is very strong--flowing out in every direction. For some years I had an underground line between my house and laboratory, and no part of the line between the two stations was above ground. Many and many times during the prevalence of a thunder-storm have the telephone-bells been made to ring at both ends of the line by a discharge from the cloud to the earth, and in some cases the discharge was several miles away. The wires could not have been affected so powerfully in any other way than through the earth.

It will be seen by the foregoing statements that it is possible to transmit messages through the earth for long distances, but the difficulty in the way of its becoming a general system is twofold. First, we cannot always have a thunder-cloud at hand from which to transmit our signals, and, secondly, the signals would be received alike at every station simultaneously.

CHAPTER XXII.

NIAGARA FALLS POWER--INTRODUCTION.

As our readers know, Niagara Falls is situated upon the Niagara River, which is the connecting-link between Lake Erie and Lake Ontario. The surface of Lake Erie lies 330 feet above that of Lake Ontario. The high level upon which Lake Erie is situated abruptly terminates at Queenstown, which is near the point where the Niagara River empties into Lake Ontario. From Lake Erie to the falls the level of the river is gradually lowered a little less than 100 feet, and most of this (making "the rapids") occurs in the last mile above the point where it takes a perpendicular plunge of 165 feet into a narrow gorge extending for seven miles, through which the river runs, gradually falling also 100 feet in that distance. The river above the falls is broad, varying from one to three miles in width, but below that point it is suddenly narrowed up to a distance of from 200 to 400 yards.

It is supposed that at one time the fall was situated at the bluff overlooking Queenstown, near Lake Ontario, and at that time was very much higher than it is at present. Through long ages of time the water has gradually eaten away the rock, thus forming the gorge. It is estimated by different geologists that the time required to wear away the rock back to the present position of the fall has required from 15,000 to 35,000 years. Some authorities place the rate of wear at three feet per annum and others not more than one. It is well known, however, that this erosion is constantly going on, and if nothing is done to check it the time will come when the gorge will extend up to Lake Erie and drain it, practically, to the bottom. This is a matter, however, that the people of this and those of several succeeding generations need not worry about.

In the early days, before the country was settled and the banks of the river were lined with trees, and no houses, hotels or horse-cars were to be seen; when the puffing of the locomotive was not heard echoing from shore to shore; when no bridges spanned the river to mar its beauty, and when nature was the only architect and beautifier, Niagara Falls must have been one of the most attractive spots on the earth; at least it is the place of all places where the mighty energies of nature are gathered together in one grand exhibition of sublime power. Here for ages this same grand exhibition had been going on, and although there was no human eye to see it, those of us who believe that nature is not a thing of chance, but that it was planned by an intelligence infinitely superior to that of any man, can easily imagine that the Great Architect and beautifier of this same nature, not only plans but enjoys the work of His own hand. Why not? For ages the same sun, in his daily round, has reflected that beautifully colored rainbow, here the product of sunshine and mist. The same water, through these successive ages, has been lifted to the clouds by the power of the sun's rays, and has been carried back to the fountain-heads on the wings of the wind, and there has been condensed into raindrops, that have fallen on land, lake and river, and in turn has been carried over this same waterfall in its onward course toward the sea, only again to be caught up into the clouds; and thus through an eternal round it has been kept moving by that mighty engine of nature, the sun. It is said that "the mill will never grind with the water that has passed." This is true only in poetry. As a matter of fact, "the water that has passed" may often return to help the mill to grind again.

Water-powers have been utilized in a small way for many years for the purpose of generating electricity through the medium of the dynamo, but nowhere in the world has the application of the force been made for this purpose on such a grand scale as at Niagara Falls. When one stands on the bank of the river and sees the great waterfall as it plunges over the precipice, exerting a force of from five to ten million horse-power, one is overwhelmed in contemplation of its possibilities as a source of energy that may be converted into work, mechanical and chemical, through the medium of electricity.

The genius of man has devised a way by which some of this constantly wasting energy may be converted into electricity and distributed to different points to perform various kinds of work. But the amount utilized as yet is scarcely a drop when compared with that which might be if the whole torrent could be set to work in the same manner as a very small portion of it now is.

CHAPTER XXIII.

NIAGARA FALLS POWER--APPLIANCES.

Some years ago a company was formed for the purpose of utilizing, to some extent, this greatest of all water-powers. A tunnel of large capacity was run from a point a short distance below the falls on a level a little above the river at that point. The general direction of this tunnel is up the river; it is about a mile and one-half in length, terminating at a point near the bank of the river a mile or more above the falls. Above the end of this tunnel an upright pit comes to the surface, where a power-house of large dimensions has been constructed of solid masonry. It is long enough at present to contain ten dynamos of mammoth size. Along the side of this power-house a deep broad canal is cut, which communicates with the river at that point, and through which flows the water that is to furnish the power. Of course the water level of this canal is the same as that of the river.

The foundations of the power-house extend to the bottom of the tunnel, which at that point is 180 feet below the surface of the ground. To put it in other words, the cellar or pit under the power-house is 180 feet deep and communicates with the great tunnel, which has its outlet below the falls.

Each of the ten dynamos is driven by a turbine water-wheel situated near the bottom of the pit heretofore described. The turbine-wheel is on the lower end of a continuous shaft, which reaches from a point near the bottom of the tunnel to a point ten or fifteen feet above the floor of the power-house (which is about on a level with the surface of the ground).

This shaft is incased in a water-tight cylinder of such diameter as will admit a sufficient amount of water, and connects with the turbine wheel at the bottom in the ordinary way. The water is admitted into the top of this cylinder from the canal, so that the wheel is under the pressure of a falling column of water over 140 feet high. The water, forcing its way out at the bottom through the turbine, revolves it and its long, upward-reaching shaft with great power, and enables it to work the dynamos in the power-house above, as will be described. The water discharges through the wheel in such a manner as to lift the whole shaft, thus taking away the tremendous end-thrust downward that would otherwise interfere greatly with the running of the machine through friction. After the water has done its work it flows off through the tunnel into the river below the falls.

To the upper end of the power-shaft is attached a great revolving umbrella-shaped hood; to the periphery (circumference) of this hood is attached a forged steel ring, 5 inches in thickness, about 12 feet in diameter and from 4 to 5 feet in width. The whole of the revolving portion--including the ring upon which are mounted the field-magnets, the hood, and the shaft running to the bottom of the pit, where the turbine wheel is attached--weighs about thirty-five tons.

The dynamos belong to the alternating type, and are comparatively simple in construction. In a previous chapter upon the dynamo it was stated that the fundamental feature was the relation that the field-magnet and the armature sustained to each other, and that in some cases the field-magnet revolves while the part that is technically called the armature remains stationary. In other cases the armature revolves and the field-magnets are stationary. In the latter case brushes and commutators are used, to catch and transfer the generated electricity, while in the former these are not needed, which simplifies the construction of the machine.

As we have stated, the dynamos used at Niagara are constructed with revolving field-magnets that are bolted on to the inner surface of the steel ring that is carried by the hood, so that there are no brushes connected with the machine except the small ones used to carry the current to the field-magnets.

The current for power purposes is generated in a large stationary armature about ten feet in diameter and of the same depth as the revolving ring. The revolutions of the ring send out currents of alternating polarity, and each of the ten machines will furnish electrical energy equal to 5000 horse-power, so that when the work that is now under way is completed 50,000 horse-power can be furnished in the form of electricity. About 35,000 horse-power is now actually delivered to the various industrial enterprises. The dynamos are set horizontally, since the shaft which connects them with the turbine wheel stands in a perpendicular position.

Not all of the energy that is developed by the water-wheel is converted into electricity, but some of it appears as heat. In order to prevent the heat from becoming so great as to be dangerous to the machine it must be constructed in such a way as to admit of sufficient ventilation for cooling purposes. The armature is so constructed that there are air-passages running all through it, and on top of the revolving hood are two bonnet-shaped air-tubes set in such a way as to force the air down through the armature, which carries off the heat and warms the power-house, on the principle of a hot-air furnace. This great machine--which, in a way, is so simple in its construction--when in action conveys to the mind of the beholder a sense of wonderful power. It is only when we stand in the presence of such exhibitions as may be seen in this power-house, devised and executed by the genius of man, and in that greater presence, the mighty Falls of Niagara, that we get something of a conception of the power of the silent yet potent energy of the great king of daylight, the sun.

There are very many interesting details that work in connection with this great power-plant, some of which we will describe, in a general way.

Standing within a few feet of each one of the great dynamos is a very beautifully constructed piece of machinery called the governor. The governor regulates the speed of the dynamos by partially opening and closing the water-gates that regulate the flow of water into the turbines. The question may be asked, why is there any regulation needed, if there is always an even head of water? There are two reasons--one because the load on the dynamo is constantly changing, and another that the head of water changes, although this latter fluctuation is in long periods. If the circuit leading out from the dynamo is broken, the rotating part of the dynamo will move with great ease and little power, as compared with what is required when the circuit is closed, and the current is going out and doing work. The increased amount of energy that will be required to keep the dynamo moving at a certain rate of speed when the load is on--in other words, when the circuit is closed--will depend upon the amount of current that is going out from the dynamo to perform work at other points. As the amount of current used outside for the various purposes is constantly changing, it follows that the load on the dynamo is constantly changing also. As the load changes, the speed will change, unless the amount of water that is flowing into the turbine is changed in a like proportion; hence the necessity for a governor that will perform this work. You can easily imagine that it will require a great amount of power to move the gate up or down with such a pressure of water behind it. It is not possible here to explain the operation of the governor in detail, as that could not be done without elaborate drawings; suffice it to say that the whole thing is controlled by a small ball governor such as we see used in ordinary steam-engines for regulating steam-pressure.

The rising or falling of the balls of this governor to only a very slight extent will bring into action a power that is driven by the turbine itself, which is able to move the water-gate in either direction according as the balls rise or fall. For instance, if the balls rise beyond their normal position, it shows that the dynamo is increasing in speed, and immediately machinery is brought into action that shuts the water off in a small degree, just enough to bring the speed back to normal. If the balls drop to any extent, it shows that the load is too great for the amount of water, and that the dynamo is decreasing in speed; immediately the power is brought into action, now in the opposite direction, and the water-gate is opened wider. These slight variations of speed are constantly going on, and the constant opening and closing of the gate follows with them. It is a beautiful piece of machinery, and is beautifully adapted to the work it has to perform. It is continually standing guard over this greater piece of machinery that is exerting an energy of 5000 horse-power and prevents it from going wrong, both in doing "that which it should not do and leaving undone that which it should do." It is a machine that, when in action, points a moral to every thinking person who beholds it. Every man has such a governor if he only has the inclination to use it.

I have said further back that the water-head varies, but usually at long periods. This variation is chiefly caused by changes of wind, and it is very much greater than one would suppose without studying the causes. Lake Erie lies in an easterly and westerly direction, and when the wind blows constantly for a time from the west, with considerable force, the water piles up at the eastern end of the lake, which causes the level of the Niagara River to rise to a very sensible extent. It is not so noticeable above the falls as below, because of the great difference in the width of the river at these two points. Sometimes the river below the falls, as it flows through the narrow gorge, will vary in height from twenty to forty feet. When the wind stops blowing from the west and suddenly changes and blows from the east, it carries the water of the lake away from the east toward the west end, which will produce a corresponding depression in the Niagara River. No doubt there is an effect produced by the difference of annual rainfall, but the effect from this cause is not so marked as that from the changing winds.

Another appliance used in the power-house, chiefly for handling heavy loads and transferring them from one point to another, is called the electric crane. It is mounted upon tracks located on each side of the power-house. The crane spans the whole distance, and runs on this track by means of trucks from one end of the power-house to the other. Running across this crane is another track which carries the lifting-machinery, consisting of block and tackle, able to sustain a weight of fifty tons. Situated at one end of the crane are one or more electric motors, which are able, under the control of the engineer, to produce a motion in any direction, which is the resultant of a compound motion of the two cars acting crosswise to each other together with the perpendicular motion of the lifting-rope connected with the block and tackle. It seems like a thing endowed with human reason, when we see it move off to a distant part of the building, reach down and pick up a piece of metal weighing several tons, carry it to some other portion of the building and lower it into place, to the fraction of an inch. While the machine itself does not reason, there is a reasoning being at the helm, who controls it and makes it subservient to his will. The machine is to the engineer who manipulates it what a man's brain is to the man himself. The brain is the instrument through which the unseen man expresses his will and impresses his work upon men and things in the visible world.

CHAPTER XXIV.

NIAGARA FALLS POWER--APPLIANCES.

In the last chapter I described some of the appliances used in connection with the power-house. There are many things that are commonplace as electrical appliances when used with currents of low voltage and small quantity, that become extremely interesting when constructed for the purpose of handling such currents as are developed by the dynamos used at Niagara. For instance, it is a very commonplace and simple thing to break and close a circuit carrying such a current as is used for ordinary telegraphic purposes, but it requires quite a complicated and scientifically constructed device to handle currents of large volume and great pressure. If such a current as is generated by a dynamo giving out 5000 horse-power under a pressure of 2200 volts should be broken at a single point in a conductor, there would be a flash and a report, attended with such a degree of heat and such power for disintegration that it would destroy the instrument.

The circuit-breakers used at Niagara are constructed with a very large number of contacts made of metal sleeves, or tubes, say one inch in diameter, so constructed that one will slide within the other; the sleeves being slotted so as to give them a little spring that secures a firm contact. These are all connected together electrically, on each half of the switch, as one conductor, so that when the switch is closed the current is divided into as many parts as there are points of contact in the switch. Suppose there are 100 of these contact-points, a one-hundredth part of the current would be flowing through each one of them. If, now, these points are so arranged that they can be all simultaneously separated, the spark that will occur at each break will be very small as compared with what it would be if the whole current were flowing through a single point, and it would be so small that there would be no danger attending the opening of the switch. These switches are carefully guarded, being boxed in and under the control of a single individual.

There is another apparatus that is a necessary part of every manufacturing or other kind of plant that uses electricity from this power-house, and this is called the transformer. Many of you are familiar with the box-shaped apparatus that is used in connection with electric lighting when the alternating current is used. Where simply heating effects are required, such as in electric lighting, for instance, the alternating current can be used to greater advantage than the direct current when it has to be carried to some distance, owing to the fact that it may be a current of high voltage. A greater amount can be carried through a small conductor; thus greatly reducing the cost of an electrical plant that distributes power to a distance. A transformer is an apparatus that changes the current from one voltage to another.

In the ordinary electric-light plant, such as is used in a small town or village, the current that is sent out from the power-station has a pressure of from 1000 to 1500 volts, according to the distance to which it is sent. It would not do, however, for the current to enter a dwelling at this high pressure, because it is dangerous to handle, and the liability to fires originating from the current would be greatly increased. At some point, therefore, outside of the building, and not a great distance from it, a transformer is inserted which changes the voltage, say, from 1000 down to 50 or 100, according to the kind of lamps used. Some lamps are constructed to be used with a current of fifty volts and others for 100 or more. The lamp must always be adapted to the current or the current to the lamp, as you choose. The human body may be placed in a circuit where such low voltage is used without danger, but it would be exceedingly dangerous to be put in contact with a pressure of 1000 or more volts, such as is used for lighting purposes.

In principle the transformer is nothing more or less than an induction-coil on a very large scale. The ordinary induction-coil, such as is used for medical purposes, is ordinarily constructed by winding a coarse wire around an iron core. This core is usually made of a bundle of soft iron wires, because the wires more readily magnetize and demagnetize than a solid iron core would. Around this coil of coarse wire, which we call the primary coil, is wound a secondary coil of finer wire. If now a battery is connected with the primary coil, which is made of the coarse wire, and the circuit is interrupted by some sort of mechanical circuit-breaker, each time the primary or battery circuit is opened there will be a momentary impulse in the secondary circuit of a much higher voltage; and at the moment the primary circuit is closed there will be another impulse in this secondary circuit in the opposite direction. The latter impulse is called the initial and the former the terminal impulse. A current created in this manner is called an induced current. The initial current is not so strong as the terminal in this particular arrangement.

If we should take hold of the two wires connected with the two poles of the battery and bring them together so as to close the circuit, and then separate them so as to break it we should scarcely feel any sensation--if there were only one or two cells, such as are ordinarily used with such coils. But if we connect these wires to the coils of the induction apparatus and then take hold of the two ends of the secondary coil and break and close the primary circuit we should feel a painful shock at each break and close, although the actual amount of current flowing through the secondary wire is not as great as that which flows through the primary; but the voltage (or electromotive force) is higher, and thus is able to drive what current there is through a conductor of higher resistance, such as the human body. For this reason there is more current forced through the body, which is a poor conductor, than can be by a direct battery current which has a lower voltage. If now we should take a battery of a number of cells, so as to get a voltage equal to that given off by the secondary coil, and connect it with the fine-wire coil instead of the coarse-wire coil--thus making what was before the secondary coil the primary--by breaking and closing the battery circuit as before we shall get a secondary or induced current in the coarse-wire coil, but it will be a current of low voltage, and will not produce the painful sensation that the secondary coil did.

We have now described the principle of a transformer as it is worked out in an ordinary induction-coil. As has been stated, at Niagara Falls the current comes from the dynamos with an electromotive force or pressure of 2200 volts. For some purposes this voltage is not high enough, and for other purposes it is too high; therefore it has to be transformed before it is used! For some purposes this transformation takes place in the power-house, and for others it takes place at the establishment where it is used. For instance, take the current that is sent to Buffalo, a distance of from twenty to thirty miles. The current first runs to a transformer connected with the power-house, where it is "stepped-up" (to use the parlance of the craft) from a voltage of 2200 to 10,000. It is carried to Buffalo through wire conductors that are strung on poles, and is there "stepped-down" again through another transformer to the voltage required for use at that place. The object of raising the voltage from 2200 to 10,000 in this case is to save money in the construction of the line of conductors between the two points. If the voltage were left at 2200--the conductors remaining the same as they are now--the loss in transmission would be very great, owing to the resistance which these wires would offer to a current of such comparatively low voltage as 2200. To overcome this difficulty--if the voltage is not increased--it would be necessary to use conductors that are very much larger in cross-section (thicker) than the present ones are. And as these conductors are made of copper the expense would be too great to admit of any profit to the company.

If we go back to an illustration we used in one of the early chapters on electricity we can better explain what takes place by increasing the voltage. If we have a column of water kept at a level say of ten feet above a hole where it discharges, that is one inch in diameter, a certain definite amount of water will discharge there each minute. If now we substitute for the hole that is one inch in diameter one that is only one-half inch in diameter a very much smaller amount of water will discharge each minute, if the head is kept at the same point--namely, ten feet. But if now we raise the column of water we shall in time reach a height which will produce a pressure that will cause as much water to discharge per minute through the one-half-inch hole as before discharged through the one-inch hole with only the pressure of a ten-foot column. This is exactly what takes place when the voltage is "stepped-up," which is equivalent to an increase of pressure.

It will be seen from the foregoing that these transformers have to be made with reference to the use the current is to be put to. In general shape they are alike in appearance, the difference being chiefly in the relation the primary sustains to the secondary coils. There is another kind of transformer that is used when it is necessary to have the current always running in the same direction. This transformer, as heretofore explained, does not change the voltage of the current, but simply transforms what was an alternating into a direct current. By alternating current we mean one that is made up of impulses of alternating polarity--first a positive and then a negative. The direct current is one whose impulses are all of one polarity. The direct current is required for all purposes where electrolysis (chemical decomposition by electricity, as of silver for silver-plating, etc.) is a part of the process. The alternating current may be used without transformation in all processes where heat is the chief factor. For motive power either current may be used, only the electromotors have to be constructed with reference to the kind of current that is used.

The rotary transformer, which may be driven by any power, consists of a wheel carrying a rotating commutator so arranged with reference to brushes that deliver the current to the commutator and carry it away from the same, that the brushes leading out from the transformer will always have impulses of the same polarity delivered to them. In the parlance of the craft, the transformers that are used to change the voltage from high to low, or vice versa, are called "static transformers," simply because they are stationary, we suppose. The others are called rotary, or moving transformers, to distinguish them from the other forms. The operation of the latter is purely mechanical, while the former is electrical. In some instances where the static transformers are very large they develop a great amount of heat, so much that it is necessary to devise means for dissipating it as fast as created. In some instances this is done by air-currents forced through them, but in others, where they are very large, oil is kept circulating through the transformer from a tank that is elevated above it, the oil being pumped back by a rotary pump into the tank where it is cooled by a coil of pipe located in the oil, through which cold water is continually circulating. By this means cold oil is constantly flowing down through the transformer, where it absorbs the heat, which in turn is pumped back into the tank, where it is cooled.

Having now traced the energy from the water-wheel through the various transformations and having described in a very general way the apparatus both for generating electricity and for transforming it to the right voltage necessary for the various uses to which it is put, we will proceed in our next chapter to follow it out to the points where it is delivered, and trace it through its processes, and the part it plays in creating the products of these various commercial establishments.

CHAPTER XXV.

ELECTRICAL PRODUCTS--CARBORUNDUM.

The production of electricity in such enormous quantities as are generated at Niagara Falls has led to many discoveries and will lead to many more. Products that at one time existed only in the chemical laboratory for experimental purposes, have been so cheapened by utilizing electrical energy in their manufacture, as to bring them into the play of every-day life. Still other products have only been discovered since the advent of heavy electrical currents. A substance called carborundum, which was discovered as late as 1891, has now become the basis of an industry of no small importance. It is a substance not unlike a diamond in hardness, and not very unlike it in its composition. The chief use to which it is put is for grinding metals and all sorts of abrasive work. It is manufactured into wheels, in structure like the emery-wheel, and serves the same purpose. It is much more expensive than the emery-wheel, but it is claimed that it will do enough more and better work to make it fully as economical.

It was my pleasure and privilege to visit the factory at Niagara Falls, and through the courtesy of Mr. Fitzgerald, the chemist in charge of the works, I learned much of the manufacture and use of carborundum. The crude materials used in the manufacture of carborundum are, sand, coke, sawdust and salt; the compound is a combination of coke and sand. It combines at a very high heat, such as can be had only from electricity. When cooled down the product forms into beautiful crystals with iridescent colors. The predominating colors are blue and green, and yet when subjected to sunlight it shows all the colors of the solar spectrum to a greater or less degree. The crystals form into hexagonal shapes, and sometimes they are quite large, from a quarter to a half inch on a side. The salt does not enter into the product as a part of the compound, neither does the sawdust. The salt acts as a flux to facilitate the union of the silica and carbon. The sawdust is put into the mixture to render it porous so that the gases that are formed by the enormous heat can readily pass off, thus preventing a dangerous explosion that might otherwise occur. In fact, these explosions have occurred, which led to the necessity of devising some means for the ready escape of the gases.

The process of manufacture as it is carried on at Niagara is interesting. The visitor is first taken into the rooms where are stored the crude material, the sand, coke, sawdust and salt. The sand is of the finest quality and very white.

The coke is first crushed and screened, the part which is reduced to sufficient fineness is mixed by machinery with the right proportion of sand, salt and sawdust. The coarser pieces of coke are used for what is called the core of the furnace, which will be described later on.

This mixture is carried to the furnace-room, which has a capacity for ten furnaces, but not all of these will be found in operation at one time. Here the workmen will be taking the manufactured material from a furnace that has been completed, and there another furnace is in process of construction, while a third is under full heat, so that one sees the whole process at a glance. These furnaces are built of brick, about sixteen feet in length and about five feet in width and depth. The ends and bed of the furnace are built of brick, and might be called stationary structures. The sides are also built of brick laid up loosely without mortar; each time the material is placed in the furnace, and each time the furnace is emptied, the side-walls are taken down.

A furnace is made ready for firing by placing a mass of the mixture on the bottom, and building the sides up about four feet high (or half the height when it shall be completed). A trough, about twenty or twenty-one inches wide and half as deep, is scooped out the whole length of the pulverized stuff, and in this is placed what has before been referred to as the core of the furnace, namely, pure coke broken into small pieces, but not pulverized, as in the case of the other mixture. The amount used is carefully weighed, so as to have the core the proper size that experiment has proved to give the best results. The core is filled in and rounded over till it is in circular form, being about twenty-one inches in diameter. At each end of the furnace the core connects with a number of carbon rods--about sixty in all--that are thirty inches long and three inches in diameter. These carbon rods are connected with a solid iron frame that stands flush with the outer end of the furnace. On the inside the spaces between the rods are packed full of graphite, which is simply carbon or coke with all the impurities driven out, so as to make good electrical connections with the core. This core corresponds, electrically speaking, to the filament in an ordinary incandescent lamp, only it is fourteen feet long and twenty-one inches in diameter. The mixed material is now piled up over this core, and the walls at the sides are built up until the whole structure stands about eight feet from the floor--a mass of the fine pulverized mixture, with a core of broken coke electrically connected at the ends. It is now ready for the application of electricity, which completes the work.

Let us go back to the transformer-room and examine the electrical appliances that bring the current down to a proper voltage to produce the heat necessary to cause a union between the silica of the sand and the carbon of the coke, which results in the beautiful carborundum crystals that we have heretofore described.

The current is delivered from the Niagara Power Company under a pressure of 2200 volts. The conductors run first into the transformer-room, which adjoins the furnace-room, and is there transformed down from 2200 volts to an average of about 200 volts. The transformers at these works have a capacity of about 1100 horse-power. About 4 per cent of this power is converted into heat in the process of transformation, making a loss in electrical energy of a little over 40 horse-power. This heat would be sufficient to destroy the transformer if some arrangement were not provided to carry it off. We have already described how this is done through the medium of a circulation of oil. Because of the low voltage and enormous quantity of the current passing from the transformer to the furnace very large conductors are required. The two conductors running to the furnace have a cross-section of eight square inches, and this enormous current, representing over 1000 horse-power, is passed through the core of the furnace, and is kept running through it constantly for a period of twenty-four to thirty-six hours.

Let us consider for a moment what 1000 horse-power means; as this will give us some conception of the enormous energy expended in producing carborundum. A horse-power is supposed to be the force that one horse can exert in pulling a load, and this is the unit of power. However, a horse-power as arbitrarily fixed is about one-quarter greater than the average real horse-power. If 1000 horses were hitched up in series, one in front of the other, and each horse should occupy the space of twelve feet, say, it would make a line of horses 12,000 feet long, which would be something over two miles. Imagine the load that a string of horses two miles long could draw, if all were pulling together, and you will get something of an idea of the energy expended during the burning of one of these carborundum furnaces.

Within a half hour after the current is turned on a gas begins to be emitted from the sides and top of the furnace, and when a match is applied to it, it lights and burns with a bluish flame during the whole process. It is estimated that over five and one-half tons of this gas is thrown off during the burning of a single furnace. This gas is called carbon monoxide, and is caused by the carbon of the coke uniting with the oxygen of the sand. When we consider the vast amount of material that comes away from the furnace in the form of gas it is easy to see why it is necessary to introduce sawdust or some equivalent material into the mixture, in order to give the whole bulk porosity, so that the gas can readily escape. We should also expect that after five and one-half tons had been taken away from the whole bulk that it would shrink in size. This is found to be the case. The top of the mass of material sinks down to a considerable extent by the end of the time it has been exposed to this intense heat. Gradually, after the current has been turned on, the core becomes heated, first to a red, and afterwards to an intense white heat. This heat is communicated to the material surrounding the core, producing various effects in the different strata, owing to the fact that it is not possible to keep a uniform heat throughout the whole bulk of material. Some of it will be "overdone" and some of it "underdone." The material which lies immediately in contact with the core will be overheated, and that, which at one stage was carborundum, has become disintegrated by overheating.

The silica of the compound has been driven off, leaving a shell of graphitic substance formed from the coke.

After the current is shut off and the furnace has cooled down, a cross-section through the whole mass becomes a very interesting study. The core itself, owing to the intense heat it has been subjected to, has had the impurities driven out of the coke, leaving a substance like black lead, that will make a mark like a lead-pencil, and is really the same substance, known as plumbago, in one of its forms. It is the carbon left after the impurities have been driven out of the coke. Surrounding the core for a distance of ten or twelve inches, radiating in every direction, beautifully colored crystals of carborundum are found, so that a single furnace will yield over 4000 pounds of this material. Beyond this point the heat has not been great enough to cause the union between the carbon and silica, which leaves a stratum of partly-formed carborundum; outside of that the mixture is found to be unchanged.

These carborundum crystals are next crushed under rollers of enormous weight, after which the crushed material is separated into various grades for use in making grinding-wheels of different degrees of fineness. This crushed material is now mixed with certain kinds of clay, to hold it together, and then pressed into wheels of various sizes in a hydraulic press, and afterward carried into kilns and burned the same as ordinary pottery or porcelain. These wheels vary in size from one to sixteen inches. The substances used as a bond in manufacturing wheels are kaolin, a kind of clay, and feldspar.

While carborundum has already a large place as a commercial product, there is no doubt but that the uses to which it will be put will vastly increase as time goes on. This product may be called an artificial one, and never would have been known had it not been for the intense heating effects that are obtained from the use of electricity. It certainly never could have been brought into play as one of the useful agencies in manufacturing and the arts. It is not known to exist as a natural product, which at first thought would seem a little strange in view of the evidences of intense heat that at one time existed in the earth. Its absence in nature is explained by Mr. Fitzgerald by the fact that "the temperatures of formation and of decomposition lie very close together."

CHAPTER XXVI.

ELECTRICAL PRODUCTS--BLEACHING-POWDER.

Another industry that has assumed large proportions at Niagara Falls, owing to the vast quantity of electricity produced there, is the manufacture of a commercial product called bleaching-powder, or chloride of lime. Every one knows that chloride of sodium is simply common salt, so extensively used wherever people and animals exist. Simple and harmless as it is, while it exists as a compound of the original elements, when separated into those elements they are each very unpleasant and even dangerous substances to handle. Salt is one of the most common substances in nature. It is found in many parts of the world in solid beds, and is one of the prominent constituents of sea-water.

Salt is a compound of chlorine and a metal called sodium. Sodium in its pure state has a strong affinity for oxygen, so much so that when a lump of it is thrown into water it takes fire and burns violently with a yellow flame. Chlorine, the substance with which it unites to form common salt, is a greenish-colored gas, the fumes of which are very offensive and very dangerous even to breathe, if the quantity is very considerable.

It is a curious fact in nature that two such substances as chlorine and sodium, both of them so difficult and dangerous to handle, should unite together to form such a useful and harmless compound as common salt. The important element in bleaching-powder is the chlorine which it contains. It is extensively used in the manufacture of paper and in all other materials where bleaching is required. The object of combining it with lime, forming a chloride of lime, is simply to have a convenient method of holding the chlorine in a safe and convenient manner until it is needed for use.

The chemical works at Niagara Falls manufacture bleaching-powder on a very large scale. The part that electricity plays is to separate the chlorine from the sodium as it exists in common salt. At the works I was first taken into a room where a large quantity of salt was stored. A belt with little carrier-buckets on it picked up this salt and carried it into another room, where it was thrown into a vast mixing-vat containing water. The salt was mixed with water until a saturated solution was obtained. In a large room, covering one-half acre or more of ground, were assembled a great number of shallow vessels, about 4 by 5 feet square and 1 foot deep. These vessels were sealed up so that they were gas-tight. Communicating with all of these vessels were pipes connecting with the great tank containing the saturated solution of salt.

From the top or cover of each vessel is a pipe running to a main pipe that carries off the chlorine gas into another room as fast as it is formed. Through each one of these vessels a current of electricity passes; the whole system consuming about 2000 horse-power. The electric current, as it passes through the brine, separates the chlorine from the sodium, the chlorine passing in the form of gas up through the pipes, before mentioned, into the main pipe, where it is carried into another large room and discharged into a system of gas-tight chambers. Upon the floor of these chambers is spread a coating of unslacked lime ground into a fine powder. The lime has a strong affinity for the chlorine gas and rapidly absorbs it, forming chloride of lime. When the lime is fully saturated with the chlorine the gas is turned off from that chamber, which is then opened up and the chloride taken out for shipment. A new coating of lime is now spread in the chamber and the gas is turned on and the process repeated.

There are a number of these chambers, so that the operation in all of its phases is going on continuously. The room where the chlorine gas is formed is thoroughly ventilated, a precaution which is very necessary in case any one of the vats should spring a leak, as they sometimes do.

In each one of these vats where the electrolytic process is going on there are two products constantly passing off; one, as before mentioned, is chlorine gas, and the other caustic soda in solution. The solution in the vat is constantly being renewed by the saturated solution of salt from the reservoir before mentioned. There is one stream continuously coming into the vat and two going out, caused by the decomposing power of the electric current. The solution of caustic soda is carried to large evaporating-pans, where the water is driven out of it, leaving the caustic soda in dry, white sticks of crystalline formation. In this process the electric current, which comes from the power-house with an energy of 2000 horse-power, has to be transformed twice; first, to bring it to the proper voltage for the work of decomposition, and, secondly, to change it from an alternating to a direct current, by which all electrolytic processes are carried on.

You will notice that the electrical energy expended in this establishment is double that used in the manufacture of carborundum.

The caustic soda, which is one of the products from the decomposition of salt, is taken to another establishment, where, by still another electrical process, metallic sodium is manufactured. The process here being a secret one, the writer did not have the privilege of examining the details.

CHAPTER XXVII

ELECTRICAL PRODUCTS--ALUMINUM.

Another comparatively new article of manufacture now produced in large quantities at Niagara Falls is aluminum. Until within the last few years this metal was not used to any extent by manufacturers, because of the great expense attending its production. Now, however, it is produced in such quantities as to make it about as cheap as brass, bulk for bulk. Aluminum is a very light metal, with a color somewhat lighter than silver; its specific gravity being about one-third that of iron. Aluminum is found in one of its compounds in great quantities in nature, especially in certain kinds of clay and in a state of silicate, as in feldspar and its associated minerals. It is found in great quantities in southern Georgia, where it is mixed with the red oxide of iron that abounds in that region. Here, it exists as alumina, which is an oxide of aluminum. Before it is taken to the reduction-works the alumina is separated from all other substances. It is a white powder, tasteless, and not easily acted upon by acids.

Electricity is the chief agent in the production of metallic aluminum. The reduction company buys this alumina, which has been separated from the clay or ores where it is mined. In a large room there are located a great number of iron vats or crucibles, lined with carbon, about two or two and one-half feet deep, five or six feet long and four feet wide.

Immediately over each vat is constructed a metal framework, through which are inserted a large number of carbon rods about eighteen or twenty inches long and from two to two and one-half inches in diameter. This framework is electrically insulated from the iron crucibles. The framework and the carbons are connected with the positive conductor of the electric current, and the vat or crucible with the negative. These conductors are very large, something like a foot in width and an inch in thickness, and made of some good conductor of electricity. They have to be very large because they carry a current equal to 3050 horse-power. The current is one of great volume, but very low voltage; the electromotive force at each vat or crucible being only about seven volts. As the process is electrolytic, and not simply a heating process, the direct current must be used, and therefore the current coming from the power-house must be transformed twice; first to bring it to a proper voltage and secondly to change it from an alternating to a direct current. These iron vats or crucibles are connected up in series, electrically, and then they are filled with the alumina and certain other materials, which act either as a flux or as a means of increasing the conductivity of the mixture; just what this substance is, is probably one of the secrets of the process. When all of the crucibles are filled with the mixture the current is turned on and is kept on continuously night and day seven days in the week. All of the material in the different crucibles is heated to redness, when the process of separation takes place. The oxygen of the alumina is thrown off as a gas, and other residuum floats to the top of the crucible and is skimmed off.

Metallic aluminum in a melted state sinks to the bottom of the crucible, where it is dipped out from time to time with large iron ladles and poured into sand and molded into blocks similar to that of pig iron. From time to time, as the metal is dipped out, fresh alumina with the other substances are thrown in on top of the crucible, so that the process is continually going on, day and night, week in and week out. The heat in the process of reducing alumina, as we have before seen, is not the chief factor; it simply serves to reduce the compound to a fluid state so that the electrolytic action can readily take place.

Therefore it is not necessary to be brought to a white heat, as it is in the case of the production of carborundum, described elsewhere.

It was extremely interesting to observe the wonderful magnetic effects that were produced in iron when brought into proximity with these enormous electrical conductors. The voltage was so low that one could handle them with impunity. The iron crucibles became so magnetic that a heavy bar of iron seven or eight feet long would cling to their sides, so that it would be held in an upright position. Bars of iron would cling to the conductor at any point along its length, and, although these conductors were carrying an energy of over 3000 horse-power, they produced no perceptible effect upon the human body. The reason for this lies in the fact, first, that the body is not made of magnetic material, and, secondly, the pressure is so low that the body--being a poor conductor--would not easily allow the low-pressure current to pass through it.

Aluminum is fast becoming an important article of commerce, and it is destined to become more and more so on account of its extreme lightness as compared to other metals.

It is found to be valuable also when used as an alloy with many of the other metals. One of the great drawbacks to its more extensive use lies in the fact that as yet no satisfactory method has been devised for soldering it. Undoubtedly in time this difficulty will be solved, when its use will be greatly increased. It is estimated that in its various compounds aluminum forms about one-twelfth of the crust of the earth.

CHAPTER XXVIII.

ELECTRICAL PRODUCTS--CALCIUM CARBIDE.

Another important use to which electricity is put at Niagara Falls is the manufacture of a new product, called calcium carbide. Like carborundum and aluminum, this product could not have been produced in commercial quantities in advance of a means for producing electricity in enormous volume.

Calcium carbide is a compound of calcium and carbon. Calcium is a white metal not found in the natural state, but exists chiefly as a carbonate of lime, which is ordinary limestone, including the various forms of marble. As a pure metal it is hard to obtain and very hard to maintain, as it readily oxidizes when in contact with the air. The symbol for calcium carbide is CaC_2, which means that a molecule of this carbide is compounded of one atom of calcium and two atoms of carbon. Ca stands for calcium and C for carbon. When the symbol has no figure following it, it means that one atom only enters into the compound; but if a figure follows, it means that as many atoms enter in as the figure represents.

The process of manufacturing calcium carbide is as follows: Ordinary lime before it is slacked is ground to a fine powder; then it is mixed with powdered coke or carbon in the proper quantities, so that when a chemical union takes place the proportion will be as before stated, one atom of calcium to two of carbon. As is well known, lime is procured by exposing ordinary limestone to a red heat for some hours together. The heat disengages the carbon dioxide, leaving only a combination of calcium and oxygen, which is common lime.

The mixture of ground lime and coke is put into a crucible that surrounds the arc of an electric light of enormous dimensions; the carbon conductors amounting to an area of one square foot or more. In order to cause the carbon to unite with the calcium a very intense heat is required, such a heat as can be obtained only in the arc of an electric light. When the enormous current is turned on (amounting to over 3000 horse-power) the mixture is melted, and after an exposure to this intense heat for a given length of time the oxygen of the unslacked lime is thrown off and the carbon unites with the calcium, which remains in the proportions of one atom of calcium to two of carbon, as before stated. This, it will be noted, is purely a heat process, and an intense one at that. No electrolytic action being required, the alternating current is used without transformation to the direct current, as is necessary in the manufacture of bleaching-powder and aluminum, both of which are electrolytic processes.

When the operation is completed the current is turned off and the compound allowed to cool. In cooling it assumes a slate color, which is slightly iridescent when exposed to light. It also crystallizes to a certain extent.

The value of this new product consists in its ability to evolve Acetylene gas in large quantities. A molecule of acetylene gas is composed of two atoms of carbon to two of hydrogen. To evolve the gas it is necessary only to pour water upon the calcium carbide, when a union takes place between the carbon of the carbide and the hydrogen of the water in the proportions above stated. If there is water enough the whole of the carbon will pass off with the gas, leaving a residuum of slacked lime.

The value of acetylene gas lies in its very intense illuminating power. This is due to the fact that the gas is very rich in carbon as compared with other illuminating gases. It burns with a pure white light when properly mixed with air or oxygen, but if there is a lack of air it burns with a smoky flame. In this case the carbon is not all consumed and escapes into the air in the form of soot or smoke, but when burned with the proper mixture of oxygen or common air it becomes one of the most brilliant of illuminants. Acetylene, like most other gases, becomes explosive when mixed with air in certain proportions. Whether it is more dangerous to handle than ordinary illuminating gases the writer is not prepared to say, as he has not had the opportunity to make a thorough comparison between it and other gases from an experimental standpoint.

Experiment, after all, is the only sure road to absolute knowledge. Theories are beautiful in books and lectures, but they often fail in the laboratory.

Acetylene is now being introduced as an illuminating gas for domestic and other purposes. Several methods of handling it have been proposed. One is to condense it into strong metal cylinders and deliver it in that form; another is to erect generators at convenient places and generate the gas as it is used.

A very ingenious contrivance has been invented for regulating the generation of the gas. A certain amount of the calcium carbide is placed in a gas-tight vessel containing water. As soon as the water comes in contact with the carbide the evolution of the gas begins. When the pressure on the inside of the vessel has reached a certain degree it is made, through mechanical contrivances, to lift the carbide out of the water and thus stop the evolution of the gas. When the pressure is relieved through the consumption of the gas at the burners it allows the carbide to drop into the water, when the evolution of the gas begins again.

Of course there is the same objection to this mode of lighting that attends all open burners; it is constantly discharging into the air the products of combustion, chiefly carbon dioxide, which is poisonous to animal life. As has been explained in some of the chapters on heat, in Volume II, the illuminating property of any gas is determined by the number of carbon particles that are contained in it, which become heated to incandescence as soon as they come in contact with the oxygen of the air, and remain so, for a brief period, during their passage between the two extremes of the flame. While acetylene equals electricity in its illuminating properties, the latter still stands without a rival when considered from a sanitary standpoint, as the use of electricity does not in any degree vitiate the air in a room where it is used.

We have now given somewhat in detail the following processes that are carried on at Niagara Falls through the agency of electricity, viz.: The reduction of aluminum from its oxide alumina; the production of the new and useful compound called carborundum; the formation of calcium carbide used for the production of acetylene gas, and a large chemical works, where bleaching-powder is made. In addition to these works, there is an establishment for the production of sodium from caustic potash, which is one of the products arising from the decomposition of salt in the bleaching-powder works. There is also another establishment for the production of phosphorus made from the bones and shells obtained from the phosphate beds that abound in some of the southern states, on the coast of the Atlantic Ocean. There is in process of construction a plant for the purpose of manufacturing chlorate of potash by an electrical process. In addition to these establishments mentioned, the electricity is furnished for power purposes to the Niagara Electric Light Company; to the electric railway between Niagara and Buffalo; to the Niagara Falls Railway, on the opposite side of the river; to the Niagara Power and Conduit Company of Buffalo, and the Niagara Development Company. This is only a small beginning of the uses to which electricity will be put as an agent for the development of heat, light and power as well as for the production of all substances where electrolysis is the chief factor. Sixteen companies or more are now using electricity from the Niagara power-house,--the whole amounting to about 35,000 horse-power.

CHAPTER XXIX.

THE NEW ERA.

When we consider the number of new products for whose existence we are indebted to electricity, and the number of old products that have heretofore existed experimentally, in the laboratory of the chemist only, that have now been brought into play as useful agents in the various arts and industries, we begin to realize that this is truly an electrical age and the dawning of a new era. How many, many things there are, familiar to the children of to-day, that were not even imagined by the children of twenty-five to fifty years ago. Fifty years ago the only useful purpose to which electricity was put was that of transmitting news from city to city by the Morse telegraphic code. It will be fifty-seven years the first of April, 1901, since the first telegraph-line was thrown open to the public. Less than thirty years ago but little advance had been made in the use of electrical appliances beyond the perfection of certain private-line instruments, and a means for multiple transmission. About twenty years ago there were evidences of the beginning of a new era in electrical development. At no time in the history of the world has wonder succeeded wonder with such rapidity, producing such astounding results that have revolutionized all our modes of doing business and all of the operations of commercial and domestic life, as during the last two decades. We set our watches by time furnished by electricity from one central point of observation. We read the tape from hour to hour, upon which is recorded the commercial pulse of the world, as it throbs in the marts of trade, by means of this same speedy messenger. We enter a street-car that is lighted and heated, and at the same time propelled by the same wonderful agent. In our homes and on our streets night is turned into day by a light that outrivals all other illuminants.

When we wish to speak to a friend who may be a mile or a thousand miles away we step to the end of a wire that comes within the walls of our dwelling and we talk to him as though face to face, and means are at hand by which we may write a letter to that same friend and deliver it to him in our own handwriting and over our own signatures, so quickly that it will appear before him in full form and completeness as soon as the last period is made at the end of the last line.

One sees, and hears, and lives more in a single day in this age of electricity and steam than he did in twelve months sixty years ago. And yet there are those who cry out against modern inventions and modern civilization, and are constantly quoting the days of their grandfathers and great-grandfathers when "life was simple" and there was "time to rest." "Why are we tormented with this thought-stimulating age?" they say. "Why are our emotions called into action by modern music and modern art? Why are we called upon to help the downtrodden and oppressed, and to help to elevate mankind to a higher level? Why cannot we be left alone in peace and quiet, to live in the easiest way?"

If this be good philosophy, then the swine, if he were a reasoning being, ought to be ranked among the greatest of philosophers--when he seeks a wallow in the sunshine and sleeps away his useless existence. If he is useful it is because some other being of a higher order uses him to help along his own existence. The man in these days who does not "keep up with the procession" is soon trodden under foot and some other man uses him as a stepping-stone to elevate himself.

Yet this is a selfish motive, after all. The world is now rapidly advancing in light, in knowledge, in power to use the infinite gifts that the Creator has hidden in nature; but hidden only to stimulate and reward our seeking. Every man can help in this grand progress,--if not by research and positive thought-power, at least by grateful acceptance and realization of what is gained. Look forward!_ As Emerson puts it: "To make habitually a new estimate--that is elevation."

###